AF678154

Managing Foreign Research and Development in the People's Republic of China

Chandos
Asian Studies Series: Contemporary Issues and Trends

Series Editor: Professor Chris Rowley,
Cass Business School, City University, UK
(email: c.rowley@city.ac.uk)

Chandos Publishing is pleased to publish this major Series of books entitled *Asian Studies: Contemporary Issues and Trends*. The Series Editor is Professor Chris Rowley, Cass Business School, City University, UK.

Asia has clearly undergone some major transformations in recent years and books in the Series examine this transformation from a number of perspectives: economic, management, social, political and cultural. We seek authors from a broad range of areas and disciplinary interests: covering, for example, business/management, political science, social science, history, sociology, gender studies, ethnography, economics and international relations, etc.

Importantly, the Series examines both current developments and possible future trends. The Series is aimed at an international market of academics and professionals working in the area. The books have been specially commissioned from leading authors. The objective is to provide the reader with an authoritative view of current thinking.

New authors: we would be delighted to hear from you if you have an idea for a book. We are interested in both shorter, practically orientated publications (45,000+ words) and longer, theoretical monographs (75,000–100,000 words). Our books can be single, joint or multi-author volumes. If you have an idea for a book, please contact the publishers or Professor Chris Rowley, the Series Editor.

Dr Glyn Jones
Chandos Publishing (Oxford) Ltd
Email: gjones@chandospublishing.com
www.chandospublishing.com

Professor Chris Rowley
Cass Business School, City University
Email: c.rowley@city.ac.uk
www.cass.city.ac.uk/faculty/c.rowley

Chandos Publishing is a privately owned and wholly independent publisher based in Oxford, UK. The aim of Chandos Publishing is to publish books of the highest possible standard: books that are both intellectually stimulating and innovative.

We are delighted and proud to count our authors from such well-known international organisations as the Asian Institute of Technology, Tsinghua University, Kookmin University, Kobe University, Kyoto Sangyo University, London School of Economics, University of Oxford, Michigan State University, Getty Research Library, University of Texas at Austin, University of South Australia, University of Newcastle, Australia, University of Melbourne, ILO, Max-Planck Institute, Duke University and the leading law firm Clifford Chance.

A key feature of Chandos Publishing's activities is the service it offers its authors and customers. Chandos Publishing recognises that its authors are at the core of its publishing ethos, and authors are treated in a friendly, efficient and timely manner. Chandos Publishing's books are marketed on an international basis, via its range of overseas agents and representatives.

Professor Chris Rowley: Dr Rowley, BA, MA (Warwick), DPhil (Nuffield College, Oxford) is Subject Group leader and the inaugural Professor of Human Resource Management at Cass Business School, City University, London, UK. He is the founding Director of the new, multi-disciplinary and internationally networked Centre for Research on Asian Management and Editor of the leading journal *Asia Pacific Business Review* (www.tandf.co.uk/journals/titles/13602381.asp). He is well known and highly regarded in the area, with visiting appointments at leading Asian universities and top journal Editorial Boards in the US and UK. He has given a range of talks and lectures to universities and companies internationally and has research and consultancy experience with unions, business and government; his previous employment includes varied work in both the public and private sectors. Professor Rowley researches in a range of areas, including international and comparative human resource management and Asia Pacific management and business. He has been awarded grants from the British Academy, an ESRC AIM International Study Fellowship and a five-year RCUK Fellowship in Asian Business and Management. He acts as a reviewer for many funding bodies, as well as for numerous journals and publishers. Professor Rowley publishes very widely, including in leading US and UK journals, with over 100 articles, 80 book chapters and other contributions and 20 edited and sole authored books.

Bulk orders: some organisations buy a number of copies of our books. If you are interested in doing this, we would be pleased to discuss a discount. Please contact Hannah Grace-Williams on email info@chandospublishing.com or telephone number +44 (0) 1993 848726.

Textbook adoptions: inspection copies are available to lecturers considering adopting a Chandos Publishing book as a textbook. Please email Hannah Grace-Williams on email info@chandospublishing.com or telephone number +44 (0) 1993 848726.

Managing Foreign Research and Development in the People's Republic of China: The new think-tank of the world

ZHENG HAN

Chandos Publishing
Oxford · England

Chandos Publishing (Oxford) Limited
TBAC Business Centre
Avenue 4
Station Lane
Witney
Oxford OX28 4BN
UK
Tel: +44 (0) 1993 848726 Fax: +44 (0) 1865 884448
Email: info@chandospublishing.com
www.chandospublishing.com

First published in Great Britain in 2008

ISBN:
978 1 84334 153 6
1 84334 153 0

© Zheng Han, 2008

British Library Cataloguing-in-Publication Data.
A catalogue record for this book is available from the British Library.

All rights reserved. No part of this publication may be reproduced, stored in or introduced into a retrieval system, or transmitted, in any form, or by any means (electronic, mechanical, photocopying, recording or otherwise) without the prior written permission of the Publishers. This publication may not be lent, resold, hired out or otherwise disposed of by way of trade in any form of binding or cover other than that in which it is published without the prior consent of the Publishers. Any person who does any unauthorised act in relation to this publication may be liable to criminal prosecution and civil claims for damages.

The Publishers make no representation, express or implied, with regard to the accuracy of the information contained in this publication and cannot accept any legal responsibility or liability for any errors or omissions.

The material contained in this publication constitutes general guidelines only and does not represent to be advice on any particular matter. No reader or purchaser should act on the basis of material contained in this publication without first taking professional advice appropriate to their particular circumstances.

Typeset in the UK by Concerto.
Printed in the UK and USA.

Contents

List of figures and tables

Figures

Tables

About the author

Zheng Han is the HANIEL Group Chair Professor in marketing and entrepreneurship at the Sino-German School of Graduate Study (CDHK), Tongji University, Shanghai, China. A native Chinese, he studied in Germany and holds a combined master's degree in mechanical engineering and business administration. He completed his PhD at the University of St Gallen in Switzerland. He was a scholar of the Swiss National Science Foundation, a visiting scholar at the China Europe International Business School (CEIBS) in China and at the Wharton-SMU Research Centre in Singapore, and a post-doctoral fellow at the Asia Research Centre of the University of St Gallen, Switzerland, and at the Jacobs Center of the Jacobs University, Bremen, Germany.

Zheng Han has gained relevant practical experience at companies including McKinsey & Co. and the research and technology division of Daimler-Chrysler in Germany. He also serves as senior adviser to a Fortune 500 company and consults for several large German companies on market-oriented innovation and China-related issues. He has published articles in journals such as *R&D Management*, *Journal of Knowledge Management* and *Journal of Change Management*.

He can be contacted at: zheng.han@unisg.ch.

List of acronyms

BLRC	Bell Labs Research China
CAS	Chinese Academy of Sciences
CCPC	China Copyright Protection Centre
CEIBS	China Europe International Business School
CEO	chief executive officer
CIETAC	China International Economic and Trade Arbitration Committee
COO	chief operating officer
CPC	Communist Party of China
CTO	chief technology officer
ECID	Economic Crimes Investigation Division
EDA	electronic design automation
EI	Engineering Index
FDI	foreign direct investment
FESCO	Beijing Foreign Enterprises Services Corporation
FIE	foreign-invested enterprise
FYP	Five-Year Plan
3G	third generation
GAC	General Administration of Customs
GDP	gross domestic product
GE	General Electric

GLORAD	Research Centre for Global R&D Management, Tsinghua University
GM	General Motors
GSK	GlaxoSmithKline
HBA	home-base-augmenting
HBE	home-base-exploiting
HEI	higher education institution
HQ	headquarters
HR	human resources
HRM	human resource management
HTDZ	high-technology development zone
IC	integrated circuit
IFC	International Finance Corporation (World Bank)
IP	intellectual property
IPR	intellectual property right
ISTP	Index to Scientific & Technical Proceedings
IT	information technology
IUC	industry-university cooperation
JV	joint venture
MBA	master of business administration
MII	Ministry of Information Industry
MNC	multinational company
MOE	Ministry of Education
MOST	Ministry of Science and Technology
NCA	National Copyright Administration
NCA	non-competition agreement
NDA	non-disclosure agreement
NDRC	National Development and Reform Commission
NERC	national engineering research centre
NPL	non-performing loan

NSFC	National Natural Science Foundation of China
OECD	Organization for Economic Cooperation and Development
PATAC	Pan Asia Technical Automotive Centre
PCT	Patent Cooperation Treaty
PPP	purchasing power parity
PR	public relations
PRB	Patent Re-examination Board
PSB	Public Security Bureau
R&D	research and development
RBV	resource-based view
RMB	renminbi (Chinese currency, also called yuan)
SAIC	Shanghai Automotive Industry Corporation Group
SAIC	State Administration of Industry and Commerce
S&T	science and technology
SAR	Special Administrative Region
SCI	Science Citation Index
SEZ	special economic zone
SIOC	Shanghai Institute of Organic Chemistry
SIPO	State Intellectual Property Office
SOE	state-owned enterprise
SSTC	State Science and Technology Commission
TCM	traditional Chinese medicine
TRIPS	Agreement on Trade-related Aspects of Intellectual Property Rights
UNCTAD	UN Conference on Trade and Development
WTO	World Trade Organization

1

Introduction

Since the late 1990s, R&D-related foreign direct investment has started to surface in emerging markets. The continued growth of the Chinese economy, together with the expansion of international investment in China, has led to a rapid increase in foreign R&D activities in the world's biggest transition economy. Besides the increasing importance of R&D internationalisation that requires new attention, foreign R&D in China was previously neglected due to the emerging status of its economy. This book, mainly based on over 100 qualitative interviews and a study of 40 multinational companies (MNCs), reveals the major motivations behind foreign R&D activities in China and the critical managerial uncertainties and challenges that MNC R&D activities face. In particular, three managerial areas comprising the most significant challenges and uncertainties are identified and investigated in more depth: managing R&D professionals in China; managing intellectual property rights; and managing cooperation with local universities. The success of foreign R&D activities in China strongly depends on the realistic estimation of its advantages and the proper identification and handling of the challenges.

The context

The rise of the People's Republic of China as a new economic power during the last decade has stunned the world. After China's opening and economic liberalisation, the GDP of the country rose at an astonishing average annual growth rate of 9.4 per cent, from US$147.3 billion to US$1.6494 trillion within the short span of 26 years, from 1978 to 2004. Accounting for one-third of global economic growth over the period 2002–2004, China has been the only leading force of global growth, aside from the USA, in the course of the recent downturn (AT Kearney, 2004). It has also become the third-largest trading nation and, according to OECD forecasts, it could become number one within five years. For leading economists, it is only a question of time before the 'Middle Kingdom', currently ranking fourth in the world in total GDP, will be as powerful as the world's economic leaders, the USA and Japan. In terms of purchasing power, China is already the second biggest economy in the world.

A major driving force of the largest transitional economy's economic renaissance is foreign direct investment (FDI). Between 1979 and 2004, China absorbed a cumulative FDI of around US$560 billion (*The Economist*, 2005a). Bolstered by China's entry into the World Trade Organization (WTO) in November 2001, FDI in China has continued to reach new heights. In 2003 China overtook the USA as the biggest recipient of FDI, which rose to US$54 billion in that year and reached an even higher level of approximately US$60 billion in 2004 and 2005 (*The Economist*, 2005a). Another 5 per cent increase occurred in 2006 (*China Daily*, 2007). This trend clearly reflects international investors' confidence in China's future economic development. A survey conducted by the UN

Conference on Trade and Development (UNCTAD) reveals that 85 per cent of the responding experts and 87 per cent of the responding MNCs consider China the most attractive global business location (UNCTAD, 2005). In general, this confidence is based upon the country's two well-known 'magnets'. First, with a population of 1.3 billion (the largest in the world), the emerging Chinese market has extremely high potential for future expansion. Second, even in the long run the country is able to offer a low production cost structure, particularly in terms of labour costs, due to the almost 'unlimited' availability of a labour force. These two reasons have logically led to a concentration of production-focused investment in China since the 1980s (The Economist Intelligence Unit, 2004). While this trend is continuously progressing, a new development is emerging in respect of foreign investment. Since the late 1990s, and especially after China's accession to the WTO in December 2001, R&D-related[1] FDI inflows in China have started to surge.[2] According to estimations by the Ministry of Commerce, MNCs' accumulated R&D investment in China reached approximately US$4 billion by June 2004, while the number of foreign-affiliated R&D centres, registered according to the eligibility criteria in place since 2000, reached 800 by the end of 2006. The expansion of foreign-invested R&D is very likely to keep increasing in future. A survey on the 'globalisation of R&D' conducted with 100 senior high-tech executives by The Economist Intelligence Unit revealed that 39 per cent favoured China as the site for future overseas R&D investments over the next three years; the USA trailed at 29 per cent and India at 28 per cent (The Economist Intelligence Unit, 2004).

Until now, most of the existing literature on R&D internationalisation has concentrated on global R&D within the triad regions – the USA, Japan and Western Europe –

since MNCs' R&D internationalisation traditionally occurs within these developed regions. Given the novelty of the trend, i.e. fast-growing global R&D activities in an emerging market like China, this area has until recently been largely ignored by further investigation. In-depth research of the phenomenon of foreign R&D in emerging markets is still rare. In particular, the body of knowledge regarding the managerial challenges of foreign R&D within the dynamic and still immature context of an emerging market such as China is underdeveloped. It is here where this book aims to contribute.

Objectives and structure

Since the research on managing foreign R&D activities in emerging markets is at such an early stage, the necessary first step of this exploratory study will be to build up a thorough understanding of the phenomenon of continually soaring R&D-related FDI in China's emerging market. Expected benefits and returns usually lie behind an investment decision. Costly R&D activities' return on investment is particularly long-term oriented, and thus of a strategic nature. The book endeavours to discover the reasons why an increasing number of MNCs have invested in R&D in China. In other words, what are the expected returns and benefits? To date, experts and scholars have identified two major driving factors for R&D internationalisation within the triad regions. While some MNCs seek to improve their existing technology and know-how assets through foreign-located R&D sites (technology-driven), others intend to promote the use of technological assets in response to specific foreign market conditions and help a company efficiently commercialise its product in foreign markets (market-driven). The question of whether MNCs set up R&D activities in China for similar motives remains unanswered. At the same time, along with potential benefits, MNCs' R&D managers will have to face unexpected managerial challenges and uncertainties within the context of an emerging market. Again, the question of the kind of uncertainties and challenges that MNCs' R&D managers have to handle in managing foreign R&D in China remains unexplored, and, even more importantly, so does the question of how these challenges can be managed and the uncertainties be overcome.

The study has been built up by two different research phases. Taking both phases together, 108 interviews were conducted with 84 interviewees and 40 MNCs were investigated. Interviews were conducted with executives and R&D managers involved in R&D activities in China and experts from consulting, university and independent research institutes. The companies interviewed are MNCs from the machinery, IT, pharmaceutical and chemical industries, such as ABB, Ciba, Lucent Technologies, Micronas, Nokia, Novartis, Schindler and Unilever, to name only a few. Depending on the industry, investment history, corporate strategy and mandate, the R&D activities of the interviewed MNCs employ from less than a dozen staff members to more than a thousand researchers (e.g. Lucent Technologies). Logically, the smaller the R&D unit, the more development- or reengineering-oriented its activities are. The majority of the companies interviewed are based in the triad regions. Many of them have a significant and traceable history in doing R&D in the country, while others have just started their R&D activities in China. This type of diversified setting helped the researcher to identify the longitudinal development of certain issues. On the other hand, this type of setting also reveals the shifting challenges during a foreign R&D centre's different stages. Besides talking to general senior R&D managers and also local R&D staff, in-depth interviews were conducted with field experts from academia, consulting and independent research institutions, local governmental officials and specialised IP attorneys. Interviewees with diversified backgrounds enable the author to dissect the topics under investigation from different perspectives and thus generate a more comprehensive insight.

The overall aspiration of research phase I was to identify the motivations for and managerial barriers to foreign R&D in China. In order to provide a contextual perspective of foreign R&D activities, Chapter 3 generates an overview and initial understanding of the emerging Chinese economy, and reveals its high potential and underlying risks. Chapter 4 provides insights into the pattern of existing foreign R&D activities in China. Based on the results of this empirical research, Chapter 5 discusses and analyses the identified motivations and challenges. The findings of Chapter 5 also lay the cornerstones for research phase II. Based on the findings of research phase I, three particular managerial areas having the most significant challenges and uncertainties for MNCs' R&D managers in China are discussed in more depth in the following chapters: managing R&D professionals in China (Chapter 6), managing intellectual property rights (Chapter 7) and managing cooperation with local universities (Chapter 8). Best practices and managerial implications in overcoming particular challenges are discussed at some length.

Notes

1 Based on the definition of the OECD, in this book research and development (R&D) activity comprises creative work 'undertaken on a systematic basis in order to increase the stock of knowledge and the use of this stock of knowledge to devise new applications' (OECD, 2002a: 30). R&D covers the activities of basic research, strategic basic research, applied research and experimental development.

2 In general, the highly concentrated global R&D expenditure grew rapidly over the past decade to reach some US$677 billion in 2002. Led by the USA, the ten biggest countries in respect of R&D expenditure account for more than four-fifths of the world's total R&D. China and the Republic of Korea are the only two developing countries among the top ten. However, the share of developed countries fell from 97 per cent in 1991 to 91 per cent in 2002, while that of developing Asia rose from 2 per cent to 6 per cent (UNCTAD, 2005).

2

Literature review

R&D is usually one of the last objectives of an internationalisation strategy (Mansfield et al., 1979; Pearce, 1989). For almost two decades, the process of R&D internationalisation has continued in earnest. For example, from 1994 to 2000 foreign-owned R&D in the USA grew at a real average annual rate of 10.8 per cent. During the same period, US-owned R&D abroad achieved an average annual growth rate of 6.9 per cent (National Science Board, 2004).

Three major research streams within the discipline of R&D internationalisation can be identified: research on the patterns of R&D internationalisation, on the motivations for R&D internationalisation and on managing foreign R&D. Existing R&D internationalisation literature provides valuable insights in respect of our understanding of the subject. This chapter will first review the existing literature in the area of R&D internationalisation, and then identify research gaps relating to the new phenomenon. Based on these gaps, research questions for the first research phase will be formulated.

Research on patterns of R&D internationalisation

Researchers such as Pearce and Pooni (1996), Kuemmerle (1999), Granstrand (1999) and von Zedtwitz and Gassmann (2002) have examined R&D internationalisation with regard to its patterns and/or process. Research in this field mainly concentrates on two subjects. First, how does the internationalisation process of R&D develop over time? Second, what is the relation between this process and the internationalisation process of other corporate functions?

The 'how' question is tackled by Cantwell (1992), for example, who examined patterns in the internationalisation process of MNCs' R&D activities and their implications for an MNC's competitiveness as well as for the country involved. The research work of Granstrand (1999) suggests that corporate R&D has generally been internationalised later than other corporate functions, such as sales and production. In fact, corporate R&D is usually at the end of the international operations' mode chain. Hence, firm internationalisation can be seen as incremental and increasing over time (Johanson and Vahlne, 1977; Granstrand, 1999) into more value-added activities. This incremental or evolutionary internationalisation process is supported by Mutinelli and Piscitello (1998) and Kuemmerle (1999) when examining the entry modes for R&D. Greenfield investments are the prevailing mode of entry, meaning that firms establish R&D sites in their home country first before taking it overseas (Kuemmerle, 1999), and that the likelihood of cooperative solutions is high when uncertainty is high and vice versa, since cooperation is effective in reducing the overall risks (Mutinelli and Piscitello, 1998).

Table 2.1 **Key literature on international R&D in the triad regions**

Country focus	Key literature
Western Europe	Sweden: Håkanson and Zander (1988), Håkanson (1992), Håkanson and Nobel (1993a, 1993b)
	UK: Pearce and Papanastassiou (1996)
	Germany: Wortmann (1990), Beise and Belitz (1996)
Japan	Westney (1993), Kenney and Florida (1994), Papanastassiou and Pearce (1994), Asakawa (1996, 2001), Odagiri and Yasuda (1996)
USA and Canada	Ronstadt (1977, 1978), Mansfield et al. (1979), Dalton and Serapio (1995), Niosi (1997)

Source: Author's research; von Zedtwitz (2004)

The literature on R&D internationalisation has mainly concentrated on international R&D within the triad regions – the USA, Japan and Western Europe. Country-focused in-depth research was conducted, among others, for the USA, Canada, Japan, Sweden and Germany (see Table 2.1).

Without a doubt, the ongoing globalisation of the world economy clearly indicates that the future trend of R&D internationalisation will go beyond the triad nations and therefore take on a new dimension. More recently, von Zedtwitz (2005) has divided different patterns of R&D internationalisation into four types (see Figure 2.1). The 'traditional' method of R&D internationalisation takes place within the triad regions (type 1). More than a decade ago, industrialised nations gradually started to invest in R&D in developing countries, or, to be precise, in emerging economies such as China and India (type 2). In order to catch up with the technologically leading nations, a small number of companies from developing countries have begun to install R&D in the triad regions (type 3). Some have even begun to conduct R&D in other emerging economies.[1]

Figure 2.1 **Patterns of R&D internationalisation**

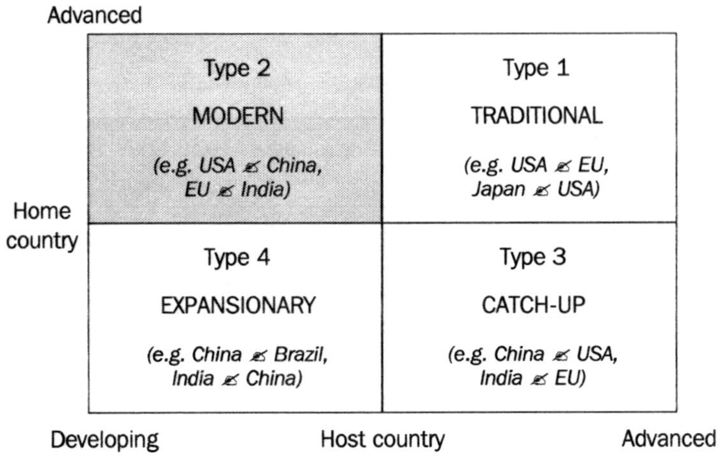

Source: von Zedtwitz (2005)

To date the 'expansionary' and 'catch-up' types of R&D internationalisation are specifically still very rare, hence not much in evidence. However, type 2 R&D internationalisation has noticeably begun to reach new dimensions in quantity and quality in recent years, particularly in China. This type of R&D internationalisation is the focal point of this research.

Research on motivations for R&D internationalisation

A number of studies have already investigated the motivation for R&D internationalisation (Ronstadt, 1977, 1978; Westney, 1993; Florida, 1997; Kuemmerle, 1999; Patel and Vega, 1999; Le Bas and Sierra, 2002). As Patel and Vega (1999) and Le Bas and Sierra (2002) indicate, adapting products and processes to suit foreign markets and providing technical support are still major elements of R&D internationalisation. In earlier studies, researchers found that the localisation of R&D is positive in helping companies to adapt their products better to the needs of local markets (Howells, 1990a; Håkanson and Nobel, 1993a). As firms establish production facilities overseas and increase product complexity, R&D facilities are needed near production sites to give technical support to localised manufacturing. Other research has found that conducting R&D overseas can help companies obtain local market knowledge (Howells, 1990b; Kogut and Chang, 1991; Florida, 1997). Moreover, companies strive to locate their R&D activities at centres of excellence and in regions that are characterised by a high rate of new technology output, therefore possessing particularly advantageous spill-over potential (Gassmann and von Zedtwitz, 1999).

Although a large percentage of the foreign-located R&D activities tend to be production-supportive, there has been evidence over the last decade of the growing significance of multinational firms seeking to harness new scientific and technological capabilities abroad. It is not surprising that the pioneers of R&D internationalisation are high-tech companies operating in small markets that face a scarcity of resources in their home countries (Gassmann and von Zedtwitz, 1999). Companies such as General Electric (GE) and General Motors (GM) in the USA, Toyota and Fujitsu in Japan and Daimler-Chrysler in Germany all had large home markets and a substantial domestic R&D base. This meant that they had less pressure to internationalise their R&D activities. However, increased competition from within and outside their industries in recent years has forced these companies to source knowledge on a global scale.

There is a common view in the studies discussed above that the motivation for R&D internationalisation can be divided into two different categories: technology-driven motivation or market-driven motivation. Kuemmerle (1999) framed these two types of international R&D investments as home-base-augmenting (HBA) and home-base-exploiting (HBE). Home-base-augmenting R&D means that firms seek to improve their existing assets, or to acquire (and internalise) or create completely new technological assets through foreign-located R&D facilities. In the case of home-base-exploiting R&D, remote R&D sites will promote the use of technological assets in response to specific foreign local conditions, helping a company to commercialise its R&D efficiently in foreign markets.

Researchers have paid much attention to why companies from industrialised countries tend to internationalise their R&D activities mostly within the triad regions. However, recent developments in R&D internationalisation have tended to take place outside these traditional regions. For instance, an increasing number of R&D sites have been created in the newly industrialised economies of South-East Asia (Gassmann and von Zedtwitz, 1999; Helble, 2004). To a certain degree, this new trend is contradictory to the technology-driven theory of R&D internationalisation, since technology progress in these emerging countries mainly lags behind that in the triad regions. Despite the increase in this phenomenon, academic research studies have so far paid little attention to the motives for doing R&D in emerging countries such as China. The question regarding the extent to which technology- and market-driven motivations have an influence on the strategy to place international R&D activities outside the triad regions has been unanswered by previous research.

Research on managing international R&D

Due to the ongoing internationalisation process of R&D, the efficiency issue regarding managing internationally dispersed R&D units has become more important (Medcof, 2001). This has led to more recent research on the management of such activities (e.g. Gassmann, 1997). International R&D organisation and management have been the subject of various research studies (e.g. Gerpott, 1990; Boutellier et al., 1996, 2000; Gassmann and von Zedtwitz, 1998; Gerybadze and Reger, 1999; von Zedtwitz, 1999; Pfeiffer, 2000; Medcof, 2001). The management of interfaces in international R&D has also gained academic attention (von Zedtwitz, 1999). Another thoroughly investigated field is the organisation of global R&D activities. Gerybadze and Reger (1999) found that transnational corporations have tended to consolidate and streamline their organisations since the mid-1990s. Distributed R&D activities and globally dispersed innovation processes have resulted in overly complex and unmanageable organisational architectures. This has induced firms to search for more efficient ways of managing their international portfolio of innovation activities. However, these studies cannot explain the phenomenon of increasing foreign R&D expansion in China. This might be due to the fact that the research samples in these studies have focused on 21 large corporations in Europe, Japan and the USA.

The management of R&D activities in an emerging market such as China is scarcely under consideration at the managerial level. There are almost no studies focusing on managerial barriers to foreign R&D in the context of a specific country outside the triad regions. Since the fast-changing Chinese environment carries high uncertainties, has barriers to doing R&D and as a country is an unfamiliar context to Western R&D managers, filling this research gap is of high importance. Extra attention and research are necessary at the managerial level in the field of the localisation of R&D in China in order to enhance the efficiency of R&D activities.

Research on R&D management in China

A comprehensive literature review has shown that R&D management in China has been largely unnoticed until recently. This can be traced back to several reasons. First, since the opening of the country, the development of the Chinese economy has been heavily based on mass production rather then cutting-edge technologies. Thus, China-based R&D has been limited in size and impact on the world. Second, the lack of recognition regarding R&D management in China is partly due to the difficulties that international scholars encounter in wishing to do adequate research in China (von Zedtwitz, 2004). Furthermore, due to the majority of domestic Chinese scholars' lack of language proficiency in the past, dialogue with the international research community was underdeveloped in this area. To a certain degree this problem still persists; however, the younger generation of management scholars in China have made constant progress in diffusing their knowledge at the international level.

One can divide literature on R&D management in China into two categories, the domestic and the international perspectives. Table 2.2 reveals several areas with a focus on

Table 2.2 Key literature on domestic R&D management in China

Area	Focus	Key literature
R&D ecosystem	National science and technology (S&T) systems	Fischer (1983), Fischer and Farr (1985), Jin and Porter (1988)
	Structural problems for the technological innovation of SOEs	Brockhoff and Guan (1996)
	China's modern national R&D capability and innovation policy	Liu and White (2001), Fischer and von Zedtwitz (2004), Huang et al. (2004)
Managerial issues	Critical issues facing domestic R&D managers	De Boer et al. (1998)
	Innovation determinants of technology ventures in China	Li and Atuahene-Gima (2001)
	Chinese domestic R&D's learning process and capabilities' development	Xie and White (2004)

domestic R&D in China that are under scholarly investigation. In the 1980s, shortly after China's economic opening, scholars primarily paid attention to the R&D ecosystem in China, i.e. the structure of national science and technology (S&T) systems and state-owned industrial R&D (Fischer, 1983; Fischer and Farr, 1985; Jin and Porter, 1988). Although their findings revealed that significant obstacles existed in the country's innovation system, China's potential for rapid advancement was considered to be high by Jin and Porter (1988). Against the background of China's increasing transformation from a centrally planned to a market-based economic system, Brockhoff and Guan (1996) examined how Chinese state-owned enterprises (SOEs) could overcome structural problems in managing technological innovation. More recently, researchers such as Liu and White (2001), Huang et al. (2004) and Fischer and von

Zedtwitz (2004) have started to investigate the country's modern R&D capability and innovation framework, and the implications for the future development of its technological capability.

Very little research has touched upon the managerial aspects of R&D in China. De Boer et al. (1998) looked at critical issues facing domestic R&D managers, such as shortage of funds, the Chinese S&T system's ageing R&D staff, increasing competition and the rapid development of technology. They suggest various solutions that could all help Chinese R&D managers to tackle critical issues, such as sending Chinese R&D managers abroad on exchanges, forging international cooperation and implementing Western R&D management theory in China. Using the example of technology ventures in China, Li and Atuahene-Gima (2001) discovered that innovation performance is linked to environmental factors, such as environmental turbulence and institutional support, and the ventures' relationship-based strategies, such as strategic alliances and political networking. As an example of advancing domestic R&D capability, Xie and White (2004) analysed the learning process and sequential capabilities' development in Lenovo, China's most successful PC manufacturer. Until now, very little knowledge has been generated in China at the R&D managerial level.

Since the late 1990s, an increasing number of MNCs have recognised China as a preferred place to conduct offshore R&D activities. As a response, some researchers have made an effort to outline the pattern and characteristics of foreign R&D investments in China (Xue and Wang, 2001; Wu, 2000; Li and Zhong, 2003), as shown in Table 2.3. From the perspective of US security policy, Walsh (2003) shed light on the nature of high-tech R&D activities under way in China, the impact these activities may have on the country's

Table 2.3 **Key literature on international R&D centres in China**

Area	Focus	Key literature
Foreign R&D in China	Characteristics of international R&D activities in China	Xue and Wang (2001), Wu (2000), Li and Zhong (2003)
	Nature of foreign high-tech R&D activities in China and their impact	Walsh (2003)
	Managing foreign R&D labs	von Zedtwitz (2004)
	Technology transfer of joint ventures	Jolly (2004)
	Motivation for and barriers to MNCs' R&D activities in China	Gassmann and Han (2004)

technological development and implications for future US-China relations. Von Zedtwitz (2004) investigated the pattern of R&D missions, geographical location, integration with parent organisations and the overall performance measurement of 199 foreign R&D units in China. He found that an increasing number of these foreign R&D laboratories are not only important vehicles for local market development, but also global competence centres with worldwide technology development mandates.

To conclude, the body of knowledge on managing foreign R&D in China is still in its infancy, considering the limited amount and scattered focuses of academic contributions. Previous research on foreign R&D in China has not been systematic, as it is a relatively new phenomenon, and thus many questions remain unanswered.

Notes

1 For example, the Chinese household appliance manufacturer Haier has established ten listening posts around the world, and

design centres in Lyons, Los Angeles, Tokyo, Amsterdam and four other cities (von Zedtwitz, 2005). The Chinese communication network solution provider Huawei had built up a central software division and institutes in Bangalore by the end of 2004 (Huawei, 2005).

3

The emerging market of China

Broadly speaking, understanding an environment should be the first attempt in approaching a subject within this environment. Elucidating the context in which international R&D is set is thus a prerequisite for a solid understanding of this phenomenon as such. Prior to investigating international R&D activities in China, this chapter will depict their context, i.e. the emerging Chinese economy. An emerging market economy is a country successfully attempting to change and improve its economy with the goal of raising itself to the level of the world's 'developed' nations.[1] These countries are often characterised by dynamic economic development accompanied by a volatile business environment and uncertainties at various levels. In the medium term, China is and will undoubtedly remain the biggest emerging market in the world.

Fast-growing economy with high potential

China is deemed one of the world's most dynamic economic powerhouses. The rise of the seventh-largest economy in the world (in terms of total GDP) began in 1978, when Deng Xiaoping, the primary architect of China's modernisation and dramatic economic development, opened China up to foreign investment and market allocation within a socialist framework. The reform has transformed the country's lethargic, unproductive, centrally planned economy into a more market-oriented system.

In a short span of 26 years, from 1978 to 2004, this 'socialism with Chinese characteristics' has increased GDP from US$147.3 billion to US$1.65 trillion, with a stunning average annual growth rate of 9.4 per cent (see Figure 3.1) (*China Daily*, 2004a). Its foreign trade rose from US$20.6 billion to US$1.15 trillion, averaging an annual growth rate of over 16 per cent. In 2004 the sum of exports and imports of goods and services amounted to around 75 per cent of China's GDP; the peak for other big economies such as Japan is 32 per cent (*The Economist*, 2004a). Besides the USA, China has been the only leading source of global growth over the course of the recent downturn, accounting for one-third of global economic growth over the period 2002–2004 (AT Kearney, 2004). Although in per capita terms the country, with 1.3 billion inhabitants, is still a

Figure 3.1 GDP development of China (1978–2004)

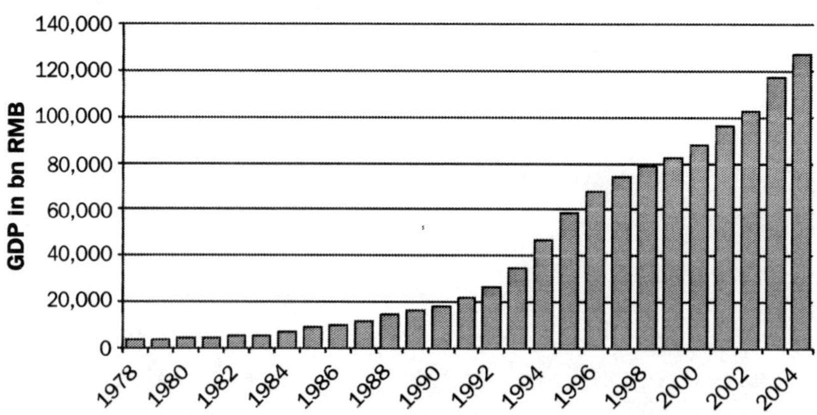

Source: National Bureau of Statistics of China (2005)

developing country, measured on a purchasing power parity (PPP) basis China is already the second-largest economy in the world after the USA.

One of the main engines pushing China's economy steadily forward is foreign direct investment (see Figure 3.2). Between 1979 and 2004 China absorbed a cumulative FDI of about US$560 billion. Underlined by China's entry into the WTO in November 2001, FDI in China has continued to reach new heights. In 2003 China overtook the USA to become the biggest recipient of FDI, rising to US$54 billion, and reaching an even higher level of an estimated US$60 billion in 2004. This trend clearly reflects the importance of the country's position as a major production site and marketplace for MNCs and their confidence in China's future economic development. In the first six months of 2005 China's foreign exchange reserves increased by around US$100 billion to US$711 billion, more than doubling from the 2003 level (*The Economist*, 2005a).

According to AT Kearney's 'FDI confidence index 2004', an annual survey of the world's largest 1,000 firms' top

Figure 3.2 **Development of FDI in China**

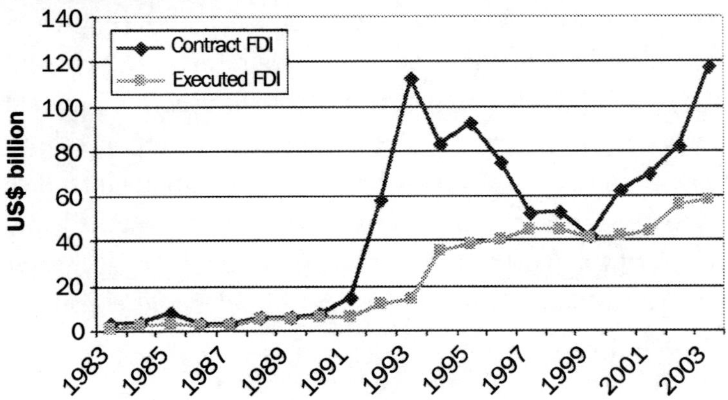

Source: National Bureau of Statistics of China (2005)

executives on their favourite FDI destinations and their investment intentions, China heads the ranking for the third consecutive year. The country is mostly favoured for its market size, growth potential, favourable cost structure and government incentives. After wholesale and retail investors, heavy and light manufacturing investors expressed the highest levels of optimism about the Chinese market (AT Kearney, 2004). For example, bypassing fixed telephone lines, China is the world's largest and most dynamic market for mobile telephony, with 235 million subscribers in 2003 and yearly growth rates of 10–15 per cent over the 2000–2003 period. Even for companies of the new economy China is a must-be place: it had 94 million internet users at the end of 2004, up from 80 million in 2003, but still only 7 per cent of the population are online. According to estimations this number could overtake America's 185 million internet users within five years (*The Economist*, 2005b).

The size of the local market is not the only reason for the growing economy. The comparatively low wage level in China is also attracting growing foreign investment. In the manufacturing industry, for example, the average wage is about US$100 per month – 5 per cent of the American average, and 10 per cent of that in some neighbouring Asian economies. Due to the unlimited labour force, mostly laid off by the agricultural sector and previously state-owned enterprises, the country will even in the long run be able to offer a competitive cost structure for labour-intensive industries. That explains why China is also called the 'factory of the world'. It is currently the largest producer of many goods such as toys, clothing, bicycles and consumer electronics; and foreign-invested enterprises produce about 45 per cent of China's total exports.

Government incentives have played a decisive role in drawing more and more foreign investors into the country since the launch of the reform. Legislation and regulations are designed to encourage foreigners to invest in high-priority sectors and regions (Walsh, 2003).[2] One of the most attractive incentives for a foreign investor is preferential tax treatment. The government grants tax relief for production-oriented foreign-invested enterprises (FIEs)[3] that invest in selected economic development zones (regional incentives) or in areas encouraged by the state (sectoral incentives), such as energy, communications and transport (UNCTAD, 2000). In general, FIEs planning to operate in China for at least ten years and engaged in production-oriented activities can enjoy an income tax exemption for two years, starting with the first profit-making year. In order to attract more technology-intensive FIEs into the country, those firms that the Ministry of Foreign Trade and Economic Cooperation has certified to be technologically advanced enterprises[4] could be granted a 50 per cent reduction in the usual income tax rate in the three consecutive years after the expiration of the initial two-year tax exemption (Income Tax Law of the People's Republic of China for enterprises with foreign investment).

Dynamic business environment with uncertainties

The astonishing speed of development and the impressive macroeconomic data brought China into the circle of the world's economic superpowers. Nonetheless, the most favoured investment destination in the world is still an emerging market. As explained in the last section, China is realising greater economic achievement through transition processes based on reforms, foreign investment and, increasingly, open markets. Simultaneously, an emerging market like China reveals greater dynamics and uncertainties compared with stable economies. Several determinants are major sources of uncertainties when doing business within this environment (see Figure 3.3). It is worth mentioning that the following consideration is based on a business ecological view rather than a managerial perspective. Determinants for micro-level uncertainties, such as cultural distance, will be discussed with respect to managing foreign R&D activities in China later in this book (see Chapter 6).

Changing macroeconomic stability

Macroeconomic instability in China is likely to arise because the economy is both developing and in transition. As an emerging economy, China faces structural changes in many areas. For example, the country is still struggling with the

Figure 3.3 **Uncertainties caused by the dynamic business environment in China**

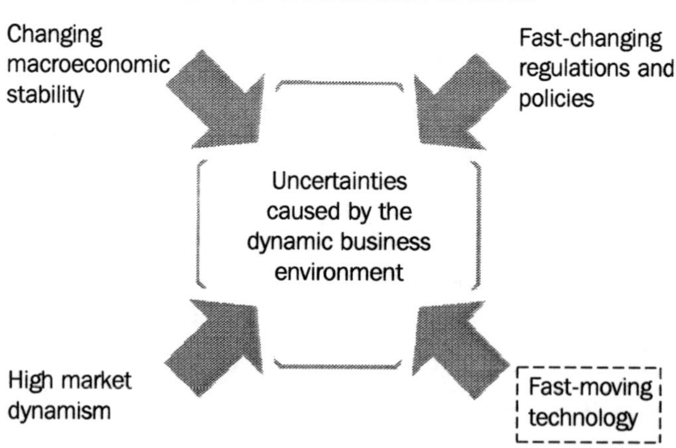

huge number of inefficient state-owned enterprises (SOEs). Although the SOEs contribute only around 25 per cent of GDP, they claim 65 per cent of total bank loans. A certain portion of these are non-performing loans (NPLs): according to estimations, NPLs account for from 22 per cent to 40–50 per cent, depending on whether the statistics come from the National Statistical Bureau or independent credit analysts (Ong, 2004). Most of the state banks are technically insolvent by any standard. Although the government's financial back-up is currently keeping the balance, any collapse of the state banks will cause the economy to implode. Thus the financial system is probably the most unstable part of the Chinese economy. In addition, most economists agree that the Chinese economy is currently overheated. There are substantial over-investments in the steel, aluminium, cement and real estate sectors, where capacity already outruns demand. Chinese policy-makers are trying to cool down the economy in order to achieve sustainable development, but it is still an open question whether the country's economy will be able to achieve a 'soft landing'.[5]

Changing regulatory environment

China's economic transformation will not be possible without the transformation of its regulatory bodies and policies. Since reform began, many small and large steps have been taken in order to accomplish progress in almost every industrial and social area. For example, the Chinese government has issued 12 policies concerning the automotive industry since 2004, which had never happened before in this industry's 50-year history. But even with this increased attention being paid to it, the fast-changing regulatory environment could make strategic planning obsolete.

The author has identified four main reasons for China's very dynamic regulatory environment. First, due to the fast-changing economic development, many of the existing regulations are out of date and need to be revised in accordance with the new conditions. Second, the fast-moving economy has also created new phenomena that authorities cannot foresee and are thus not prepared for. The result is often a vacuum or grey regulation zones. In some cases, authorities are overstrained by issues they have never dealt with before. Reactions towards new phenomena are often simply learning-by-doing processes. Regular updates are thus quite usual. Third, the Chinese government is very eager to learn from leading countries. At the same time, due to the development's evolving status, the country's indigenous characteristics and political reasons, many well-tested Western models cannot be introduced without modification, thus some regulatory changes are made through various intermediate steps. Fourth, in general, Chinese officials face much greater complexity of decision-making than their colleagues in the West. On the one hand, new regulations should support the country's further development. On the other hand, officials should also take the large development gaps in the country into consideration. Furthermore, the political interests of central and local governments should also be properly reflected.

High market dynamism

There are several reasons for China's high market dynamism. It is, for example, less easy to forecast customer demand in an emerging market. The retail market serves as an example: it has been evolving steadily over the past two decades, but with the middle class rapidly increasing and an overall increased standard of living, consumer expectations can change rapidly. Consumers who could not afford certain products due to a lack of purchasing power can become a very interesting target group for similar products within a comparatively short period. In addition, consumers' expectations of general product quality have been steadily rising, owing to the increase in income and indigenous sociological forces, which have prompted high expectations of a better lifestyle.

While the consumer demand in an emerging market is difficult to predict, new competitors keep arriving on the scene and bring new services and products, making competition fiercer and the market more dynamic and sophisticated. For instance, until 1998 there were only three multinational mobile phone manufacturers (Ericsson, Nokia and Motorola) and no domestic mobile phone manufacturer at all. By 2002 there were 27 manufacturers in the Chinese mobile phone market, with 13 domestic manufacturers having a total market share of more than 44 per cent in 2005 (*People's Daily*, 2005a).

Fast-changing technological environment

As discussed above, the business environment of an emerging market is determined by macroeconomic changes, changing governmental regulations and policies and high market dynamism. One further source of uncertainty within an emerging market is the fast-changing technological environment – of particular importance for technology-intensive companies. When talking about emerging markets or developing countries one may not always link them with cutting-edge technologies. However, specific emerging markets in the world, such as China, are already top-tier players in some disciplines' technological advancement, such

Figure 3.4 **Soaring patents and inventions in China (1993–2003)**

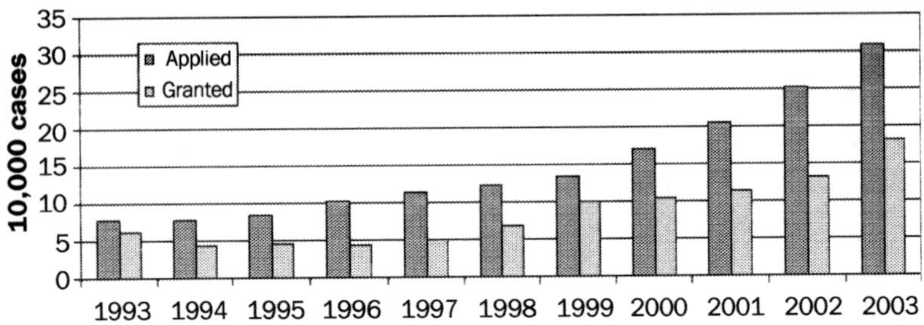

Source: China Science and Technology Statistics (2004)

as information technology or gene research. In the case of China, MNCs are aids to help the country climb the high-tech ladder, as a large portion of technology was transferred into the country by MNCs after its opening. Then again, most of these transferred technologies belong to past generations. Over the years, due to intentional government regulations, a soaring number of competitors and fiercer competition, such technological gaps in comparison to the world level are becoming smaller and smaller. To date, not many MNCs can afford to compete in the Chinese market without using their cutting-edge technologies. While MNCs have increased the speed of their technology transfer to China, domestic efforts in certain 'pillar industries', no doubt with the government's strong support, have also impelled the technological advancement of local companies. An example of this is that the total number of domestic patents and inventions in China has increased steadily (see Figure 3.4). The dynamism of the Chinese technological environment is increasingly in accordance with accelerating global technological development.

Notes

1 The term 'emerging market economy' was first coined by Antoine W. van Agtmael of the International Finance Corporation (IFC) of the World Bank in 1981. The IFC has defined an emerging market as a stock market that is in transition: increasing in size, activity or level of sophistication (Heakal, 2003).

2 For example, over time the government has eliminated time restrictions on the establishment of joint ventures, allowed foreign partners to become chairs of joint venture boards and allowed the formation of wholly foreign-owned enterprises in certain industry sectors.

3 A foreign-invested enterprise (FIE) is defined as a Chinese-foreign equity joint venture, a Chinese-foreign cooperative joint venture or a wholly foreign-owned enterprise established in China (UNCTAD, 2000).

4 A technologically advanced enterprise must possess technologically advanced production techniques and equipment, and these techniques and equipment must either be in short supply in China or the enterprise must develop new products, products that replace existing domestic products or products that will expand exports or serve as import substitutes (UNCTAD, 2000).

5 In developed economies a hard landing implies negative growth, but not in China, where growth has averaged 9 per cent over the past two decades. A soft landing would be growth slowing from its current 10 per cent to not much less than 7 per cent, the minimum needed to create enough jobs to absorb surplus rural labour and workers laid off by state-owned firms. Even that could still imply a halving in the rates of growth in industrial production and investment. A hard landing means growth significantly below 7 per cent (*The Economist*, 2004b).

4

Patterns of foreign R&D in China

At the early stages of China's economic liberalisation, production-based FDI was the main focus driver of China's economic growth and modernisation. At that time, foreign firms investing in China engaged in only limited levels of R&D activity. According to an OECD (2002b) report, only 1 per cent of foreign companies in China had R&D departments, half of these did not receive stable funding, one-third did not perform R&D regularly and nearly two in five lacked the necessary experimentation and testing equipment. A study by the Ministry of Science and Technology in 2003 provided clear evidence of longitudinal increase. Between 1987 and 2002 65 multinational companies established 82 R&D organisations (equity-based) in China (China Science and Technology Statistics, 2003). In recent years this trend has grown steadily; between 1988 and 1992 only an average of 0.8 new R&D units were established in China per year, while between 1998 and 2002 this number had grown to 8.4 units. These statistics might not reveal the exact scale of foreign R&D activities in China (e.g. Motorola, which has 18 R&D centres and seven more under construction, is only counted once in the statistics). The Ministry of Commerce's latest statistics (Xinhua News Agency, 2005) reveal more than 700 foreign R&D centres in

China by June 2005. Around 300 R&D centres are said to have been established in China in the first nine months of 2004 alone.

Several development trends in China have accelerated the expansion of foreign R&D. First, as mentioned by von Zedtwitz (2004), multinationals with previous experience in running manufacturing operations have accumulated valuable knowledge and experience in dealing with the Chinese environment and other parties, such as local partners and governmental authorities. This learning process provides them with more confidence to expand their value chain further in the country. Second, new opportunities for business and sustainable investment have emerged through continuing reform and structural changes in China's policy and economy, such as China's entry into the WTO (von Zedtwitz, 2004). Third, as the Chinese market has become more open to the world and certain industries are less controlled by the government, competition has been steadily reaching higher levels.

Although foreign R&D investments in China grow continuously, to date there is no specific Chinese governmental agency officially in charge of systematically and regularly tracking international R&D activities in China. Sporadically, there are reports from governmental organisations.[1] This lack of reliable statistics is a common problem of China's transitional statistical system, and some local governments therefore collect their own data. Whether these data will be centralised and analysed is an open question. Much available information and many studies in this area have been provided by academic institutions (e.g. GLORAD[2]), press agencies (Xinhua News Agency[3]) and consulting companies (e.g. Greatwall Strategy Consultants). Based on available public data, reports, news archives and literature, this chapter will portray the international state-of-the-art R&D investments in China by means of triangulation, and thus answer the following questions.

- Which types of international R&D activities are there in China, and what are their corporate missions?
- What is the current extent of these international R&D activities, and where are they from?
- Where are they located in China, and to which industries do they belong?

Types and missions

There are three main ways for MNCs to implement R&D activities in China. R&D departments within MNCs' China investment, which includes joint ventures, carry out the majority of foreign R&D activities. In general, this kind of activity relies on the organic growth of MNCs' Chinese investment. According to the 2002 survey by China Science and Technology Statistics (2003), 65.9 per cent of R&D centres are embedded in MNC subsidiaries in China, while the remainder of the MNCs' R&D sites are registered as wholly owned legal entities. The same study disclosed that the tendency towards wholly owned R&D centres is positively related to the technological sensitivity of MNCs' business fields, since wholly owned R&D centres offer higher protection of knowledge and prevent unwanted technology transfers; this is supported by findings from the present interviews. R&D joint ventures appear more often in industries where state control remains strong, e.g. the automobile industry. Nissan, for example, is currently building a US$40 million joint R&D lab with Dongfeng Motors in Guangzhou; General Motors established the Pan

Asia Technical Automotive Centre (PATAC), a US$50 million R&D joint venture with their local partner Shanghai Automotive Industry Corporation Group (SAIC), in 1997. The third way of carrying out R&D in China is through cooperation with local universities and research institutes. However, this kind of activity usually complements the other two and often serves as a means to 'test the water' for future R&D investments (see also Chapter 8).

Greatwall Strategy Consultants' report (2002) depicts three kinds of missions which foreign firms pursue with their local R&D sites: first, foreign R&D units serve as a linkage between China's specific market demand and know-how and their parent company's technology; second, these R&D units are important elements of corporate R&D; third, localised R&D activities also demonstrate the companies' commitment to China's economy. As stated by von Zedtwitz (2004), most foreign R&D endeavours have clearly delineated specific missions predetermined by their parent companies. For some international R&D centres, their mission goes beyond solely supporting the company's local activities: they have important charters within the global R&D network of the parent MNC. For instance, Nokia shifted a significant part of its third-generation software development to Hangzhou, transferring technologies and people from the former competence centre in Finland. The author's field research has uncovered that some foreign R&D centres do not carry out R&D as commonly defined, but rather focus on technology monitoring and corporate R&D representation. Some of them may even have public relations missions, as Chinese policy-makers prefer to see MNCs willing to establish their technology in the country by means of creating an 'R&D centre'. However, they 'often form the nucleus of more significant future R&D investment' (von Zedtwitz, 2004: 443).

Research by Li and Zhong (2003) revealed that between 1995 and 2000 two-thirds of MNCs' R&D alliances in China (including equity-based and non-equity-based alliances) were development-oriented. This finding also applied to R&D centres. To date, international R&D centres' emphasis clearly lies on development rather than on research. However, an increasing number of foreign companies have revealed their desire to extend their functional decentralised research in China. Several high-tech multinationals in the IT, pharmaceutical and telecommunications industries (e.g. Sun Microsystems, Nokia, Roche, Novartis) have added adequate resources to build specific leading-edge platforms, transforming the lab into a competency centre dedicated to fundamental research for the entire global enterprise, For instance, Microsoft set up its research centre in Beijing in 1998, and is conducting research on topics such as next-generation multimedia and human-machine-interface technology (Gelb, 2000).

Number and origins

Starting in 2000, different researchers, organisations and governmental institutions have endeavoured to portray the scale of international R&D activities established in China. Estimates vary on the actual number of foreign-invested R&D centres in China. Despite different geographical and industrial scopes and individual definitions of foreign R&D activities in each investigation or estimation, the overall trend is very clear. As Table 4.1 shows, in the period 2000–2005 the number of foreign R&D facilities grew rapidly.

Table 4.1 Reports on the number of foreign R&D centres in China

Number	Focus	Year	Source
34	China	Early 2000	Xue and Wang (2001)
40	Shanghai	12/2001	Xinhua News Agency (2001)
80	Shanghai	02/2003	Xinhua News Agency (2003)
82	China	08/2002	China S&T Statistics (2003)
100	China	2002	Greatwall Strategy Consultants (2002)
120	China	06/2002	*People's Daily* (2002a)
148	China	2002	*Asia Pulse* (2002)
223	ICT companies in China	2002	Walsh (2003)
400	China	10/2002	*People's Daily* (2002b)
600	China	08/2004	*People's Daily* (2004a)
700	China	06/2005	Xinhua News Agency (2005)
750	China	2006	Based on author's estimation

Source: Author; von Zedtwitz (2004)

At the beginning of the new millennium, only a few dozen foreign R&D centres were known in China. However, as mentioned earlier, by the end of June 2005 around 700 foreign R&D centres had been set up in China in various forms (Xinhua News Agency, 2005). According to the China Economic Review *Foreign Investment Report*, 11 MNCs announced new plans for R&D activities in five different cities in May and June 2005 alone (China Economic Review, 2005). Based on the latest available figures, the author's own research suggests that by October 2005 at least 750 foreign R&D centres or units had already been established in China or would be implemented before the end of 2006. Many MNCs have announced further strengthening of their R&D activities in China in terms of scope as well as scale.

MNCs with R&D in China usually come from the triad regions. The USA, the European Union, Japan and the Virgin Islands are major sources of R&D funds. A further significant group of R&D investors come from Greater China, specifically Hong Kong SAR and Taiwan (Li and Zhong, 2003; China Science and Technology Statistics, 2003; *People's Daily*, 2004a). By 2004 a total of more than US$4 billion had been invested in R&D in the electronics, telecommunications, transportation, medicine, chemical material and production industries. The computer and telecommunications industries are driving R&D investment in China. Microsoft ranks number one, with a US$750 million investment, followed by Lucent (US$400 million), Alcatel (US$150 million) and Motorola (US$90 million) (Zhang, 2005).

Locations

The regional distribution of foreign R&D activities reveals a clear concentration in the two most economically important cities in China, Beijing and Shanghai (*People's Daily*, 2004a). Table 4.2 reveals the multiple factors that influence the location decision. In terms of concentration of MNC R&D centres, Beijing and Shanghai may become the leading cities in the world. Partly due to established industry structures, American firms tend to go to Beijing and European firms to Shanghai. Often existing investment, especially in manufacturing, decisively influences the location decision for R&D sites. In some cases an R&D unit emerges through an initial attachment to a local representative office before it branches off in the vicinity. While IT, telecommunications and electronics companies

Table 4.2 Influencing factors for location decisions

Influencing factors	Important if...
Existing investment	R&D centres aim mainly to support existing internal customers (crucial for development activities)
Local customer	Local customers' needs strongly influence the R&D of products (crucial for development activities)
Governmental agencies	The industry strongly depends on governmental regulations
Scientific knowledge	R&D centres need closer linkage with local scientific communities, i.e. universities, state research institutes and high-tech zones (crucial for research activities)
Availability of HR	Highly qualified engineers and expatriate managers are frequently needed

prefer Beijing as an R&D site, Shanghai attracts more food, pharmaceutical and engineering companies (Greatwall Strategy Consultants, 2002; von Zedtwitz, 2004).

By the end of 2004 189 foreign R&D centres were located in Beijing alone, with almost 60 per cent of them in the ICT industry (UNCTAD, 2005). In line with von Zedtwitz's (2004) findings, the interview results indicate that for those industries in which governmental control is strong (e.g. telecommunications), Beijing is preferable. As the capital of China, Beijing is 'home to most standard-setting and decision-shaping bodies' (von Zedtwitz, 2004: 446). Networking and access to key organisations and individuals enable access to first-hand information, or even direct or indirect influence on political decisions regarding regulations and technology standards valid for the Chinese market as a whole. Furthermore, the highest concentration of engineering schools in the country and almost 100 universities make Beijing the country's unchallenged academic centre. Consequently, the availability of highly qualified R&D staff is secured. In addition, proximity to China's knowledge hubs, i.e. the Chinese Academy of Sciences, leading universities such as Tsinghua and Beijing Universities and high-tech zones such as the Zhong Guang Cun area, further underlines Beijing's advantage for research-intensive R&D units.

According to UNCTAD (2005), over 140 MNC R&D centres had been established in Shanghai by the end of 2004, of which 91 are in the Pudong New District. Shanghai is China's main centre for industry, commerce and foreign trade, and is the most dynamic and complex market in China. As the leading pacesetter in the Chinese economy, Shanghai traditionally hosts a large share of foreign investment headquarters; thus many MNC R&D units have chosen the city due to its proximity to their parent company. In general, the Shanghai environment is more business-driven. For industries where style, fashion or taste is critical, Shanghai is the preferred choice for R&D (von Zedtwitz, 2004). There are also more development-oriented centres in Shanghai and surrounding cities (China Science and Technology Statistics, 2003). It is worth mentioning that Shanghai is home to several of the nation's best schools, such as Fudan, Tongji, Jiaotong University and the China Europe International Business School (CEIBS).

Nonetheless, to a certain degree distance is also appreciated by some MNCs. As more Chinese cities and regions are trying to capture the attention of MNCs through various incentives, more MNCs have invested outside the two established hubs of Beijing and Shanghai. A significant number of newly established R&D centres have been founded in the so-called second-tier cities. As statistics show, 89 per cent of foreign R&D sites are located along a relatively short (given China's geographic size) strip between

Beijing and Shanghai (von Zedtwitz, 2004). Cities such as Tianjin (near Beijing), Suzhou (near Shanghai) and Nanjing are only a short drive from the main city centres but are not yet overcrowded with foreign companies' competitors. There is generally less competition with other foreign companies and lower employee turnover rates can be expected for companies that locate to second-tier cities. In addition, relations with Chinese government officials are easier to establish in such cities and are closer. In 2001 Nokia established a new R&D centre in Hangzhou, a city near Shanghai (200 km), and recently announced plans to consolidate its four existing joint ventures there. Motorola has established a new integrated circuit centre in Suzhou, in eastern Jiangsu province (90 km from Shanghai), where Microsoft also has its third R&D centre in China after Beijing and Shanghai. In 2005 Nokia established an R&D centre in Chengdu that focuses on mobile applications based on 3G and IP multimedia subsystems (Nokia, 2005).

This chapter has outlined the current dimension of international R&D in China. Within the last five years, the amount of international R&D activities has increased almost ten times. What are the driving motivations behind this dramatic trend? What are the major managerial challenges that international R&D faces in China? The next chapter aims to provide answers to these two questions.

Notes

1 An official report by China Science and Technology Statistics in 2003 mentioned 82 international R&D units in China: this most certainly did not capture the whole picture (see China Science and Technology Statistics, 2003). A report by the Ministry of Commerce at the end of June 2005 reveals that

approximately 700 foreign R&D centres have been set up in various forms in China (see Xinhua News Agency, 2005).

2 Headed by Professors Maximilian von Zedtwitz and Oliver Gassmann, GLORAD (Research Centre for Global R&D Management) at Tsinghua University in Beijing probably maintains the most comprehensive database on international R&D activities in China.

3 Xinhua News Agency is the Chinese national news agency.

5

Motivations and challenges of foreign R&D in China

In this chapter, the findings of the first research phase will be discussed in depth. (Parts of the findings have been published in *R&D Management*'s special issue on R&D in China (Gassmann and Han, 2004) and also cited in UNCTAD's (2005) *World Investment Report.*) First of all, two practical examples of MNC R&D activities in China, Unilever Research China and Siemens VDO Automotive, are introduced. The following sections outline the research findings, explaining the reasons why international companies have started investing heavily in R&D in China, and discussing what substantial challenges foreign R&D activities in China face. The concluding section recapitulates the most significant findings.

The case of Unilever Research China

As a world-leading supplier of consumer goods, Unilever offers about 400 products in the areas of food, home and personal care in over 150 countries. In 2004 Unilever achieved a turnover of US$48.4 billion with the help of its 223,000 employees worldwide. Unilever started to invest seriously in China from 1986 onwards. Its investment, in excess of US$1 billion since 1986, reflects the strategic importance of China for Unilever as one of the largest consumer markets. A few years after the initial investment, Unilever recognised that the technical capability in areas of quality control was insufficient to achieve its long-term goal. Products for the Chinese market that had been transferred directly from abroad rarely succeeded or met the needs of local customers. Product adaptation and development capabilities were thus gradually built up in Unilever's different joint ventures within its China investments. But since the beginning of the 1990s Unilever has increasingly recognised that the potential of doing R&D in China may lie beyond simple product adaptation. The high quality of Chinese scientists and science graduates has allowed the decision-makers to understand the strategic significance of doing R&D in the country. China as an R&D site could even become a full member of Unilever's global R&D network.

Considered as 'testing the water', a joint lab for synthetic research was founded at the Shanghai Institute of Organic Chemistry (SIOC), an institution of the Chinese Academy of Sciences (CAS), in 1996. It is an R&D joint venture that focuses on synthetic materials, a scientific area that has been strongly developed by Chinese scientists. While the CAS provided R&D staff and the building, Unilever provided the funding and the research topic, and also leads the project management. Motivated by the untapped potential of traditional Chinese medicine (TCM), an internal research group focusing on TCM and natural products was set up at Unilever's facility in Minhang, a periphery of the city of Shanghai, in 1999. Representing one of Unilever's core R&D capabilities, this group works on the exploration of unknown, natural TCM compounds. The use of TCM herbal remedies' multiple active compounds could enable the creation of novel and healthy products for the food and personal-care business with greater customer benefits and market value.

Previous experience provided solid foundations for Unilever's R&D activities' new platform in China. In October 2003 Unilever opened Unilever Research China with an initial investment of RMB166 million, after three years of preparation. Based on previous R&D activities, the new centre has core R&D capabilities in the areas of synthetic materials, micro-structured particles (nano-technology) and TCM. The new R&D hub also supports the local and regional business by generating insights regarding Chinese consumers and anticipating their future needs, identifying and driving new business opportunities and providing the manufacturing industry with technical support. As the sixth largest of the company's global R&D laboratories, Unilever Research China now employs a team of 150 R&D professionals, of whom more than 50 per cent are involved in basic research. Most are young postgraduates with qualifications and expertise in the fields of chemistry, photochemistry, material science, molecular biology and microbiology, and more than half are female scientists. Most have overseas study and work experience. Furthermore, through cooperation efforts, a large number of external leading scientists from universities and governmental institutes, such as Fudan, Peking, Jiaotong and Shanghai Engineering and Technology, are working with Unilever scientists on different programmes.

Based on the in-depth interviews, several of Unilever's key motivations for implementing R&D in China can be summarised. At the beginning of Unilever's undertaking in China, development capability was nurtured in order to fulfil the company's ambitious growth goal in the country and meet the needs of Chinese consumers. Later on, Unilever started to explore scientific achievements in China through local R&D, recognising the high quality of the local scientific work. With consumers increasingly emphasising healthy products, TCM know-how generated by Unilever's China labs could even contribute to the company's global success. Unilever Research China has built up an analysis and testing lab acknowledged company-wide. As an internal service provider, it is also well known for its world-class analysis specialists who operate at a very competitive cost level.

The case of Siemens VDO Automotive

Siemens has a corporate research lab in Beijing, employing 25 people. The mission is to contribute to corporate research (active research), to handle Siemens' intellectual property issues in China (local IP presence) and to promote Siemens standards in the early and emerging phase before dominant designs have been established (market development). In addition to this corporate R&D activity, Siemens VDO Automotive, a wholly owned subsidiary, is in the process of establishing a development centre in China.

Siemens VDO Automotive has 137 business locations worldwide with close to 43,600 employees in the all-important automotive market. Siemens VDO is a supplier of automotive products, with 16 product divisions that can be considered as profit centres, all embedded in a matrix organisation. More than 70 per cent of the product range is less than three years old, which highlights the emphasis on innovation. The Asia strategy of Siemens VDO is to expand its position and double turnover in the region by 2005/2006.

China is considered to be a strategically important market in achieving this goal. In China, Siemens VDO started its activities in 1995, and has since invested in five operating units. In the 1990s, establishing joint ventures was mandatory in order to enter the Chinese market. Since that time, two of the units have been turned into wholly owned enterprises. Siemens VDO's local production facilities in China mostly manufacture products that were developed in Europe. The majority of its customers are international auto manufacturers such as Volkswagen, BMW and Hyundai. Until now, all R&D projects have been development- and application-oriented, with no research activities at all. Due to the lack of local development capacity, these projects are supported and staffed by members of R&D teams based in other countries, such as Korea and Australia.

The management of Siemens VDO is aware of the importance of the Chinese market and the strategic significance of creating competitive advantages in China. For Siemens, there are several reasons to carry out local development. First, with new models continually being introduced into the Chinese car market, it is vital to keep product development in line with model releases from different car manufacturers. In order to respond quickly to customer demands in future, it will not be sufficient to conduct development activities through the headquarters R&D department. Second, there is cost-saving potential due to the low engineering man-hour costs in China. Cost savings can be achieved in products developed abroad (e.g. in Germany), which are sometimes over-engineered with regard to local customers' fewer requirements. Third, engineering support to local customers is easier to provide than support through a central R&D department. Fourth, due to the specific local conditions in which automotive products operate, some of the development activities must be conducted on-site. Examples of such activities include local air and climatic conditions, fuel quality needs and development work in power-train adjustment. On-board infotainment services also need localised adjustments with respect to culture and language.

At this stage, the volume of products that could be developed locally is still inadequate to achieve sufficient economies of scale. One major reason for this is the congested car market in China. About 2,400 companies are involved in the manufacturing of vehicles. At present there are 120 complete vehicle plants, of which just 12 have a yearly output exceeding 10,000 units and only three have a production capacity exceeding 300,000. According to industry experts, an enterprise with an annual output of less than 1 million is not competitive in the international market. One further dilemma that is a general concern for all automotive R&D managers in China is the lack of experienced auto specialists. Although there are several universities in China with automotive departments, there are still insufficient numbers of qualified automotive engineers given the increasing demand for specialists in the country.

The management of Siemens VDO is aware that the aforementioned obstacles set limitations on direct investment in R&D activities. Cooperation is therefore considered an alternative means to ensure competitive advantage. It is notable that most of the domestic automotive companies are not able to provide Siemens VDO with technological benefit through their relationship; within this industry, cooperation in China mainly offers advantages with respect to development time reduction, joint use of resources and facilities, regional market knowledge and market barriers to entry. In this case study there are two kinds of potential cooperation partners: universities and automotive companies along the supply chain. In the

Siemens VDO case, universities are preferable cooperation partners, primarily for two reasons: first, contact with a university is considered a good source of qualified automotive engineers; and second, universities might possess testing facilities crucial for product development, such as an engine testing facility. Both reasons are resource-based. Siemens VDO has a strong exchange and cooperation agreement with the automotive faculty of Changchun University of Science and Technology. Another potential source of cooperation for Siemens VDO is Chinese automotive firms along the supply chain (vertical integration). They could help to attract customers and build long-term business relationships with key customers in order to create market entry barriers. According to the interviewed experts, competitors will definitely not be considered cooperation partners due to the fear of undesired technology transfer.

Motivations

Prior to examining MNCs' motivations for establishing R&D in China, the results of earlier research on the basic motivations and drivers for conducting R&D abroad will be briefly outlined. Different approaches have been used to classify motivations for R&D internationalisation. One approach broadly distinguishes between demand-oriented and supply-oriented drivers (Granstrand et al., 1993; Dunning and Narula, 1995; OECD, 1998). Demand-oriented motivation factors include the local country/market's special needs that require modifications of the firm's products, or host-country restrictions, such as local content requirements, tolls, import quotas and fulfilment of standards. Supply-oriented factors include highly sophisticated foreign scientific infrastructure (e.g. new regional technological competence centres such as Silicon Valley, Prato or Modena) that takes advantage of the host country's scientific and knowledge inputs and accesses cutting-edge technology. The availability of well-educated local R&D specialists, ideally combined with low R&D personnel costs, is a further supply-oriented incentive to establish R&D abroad. A third group of drivers, environmental motivational factors, are mentioned by Granstrand et al. (1992).

In a more refined classification scheme, Beckmann and Fischer (1994) identified five categories of R&D internationalisation drivers (input-oriented, output-oriented, external, efficiency-oriented and political/social-cultural). Input- and output-oriented factors are, in principle, in accordance with the supply- and demand-oriented view. The three other categories of external, efficiency-oriented and political/social-cultural factors reveal the multifarious character of motivations for R&D internationalisation.

Based on the literature analysis, this research has relied on the Beckmann and Fischer (1994) scheme as a preliminary framework within which to examine the specific motivations for establishing R&D in the Chinese context. Since the motivation factors identified within the Chinese context do not expediently conform to Beckmann and Fischer's scheme, this research has modified the scheme by merging output- and efficiency-oriented motivational factors with performance-oriented factors, and external- and political/social-cultural factors with business-ecological factors. Table 5.1 illustrates the main motivational factors for foreign R&D endeavours in China. In the following sections these identified motivations are discussed in more detail.

Table 5.1 Identified motivations for foreign R&D activities in China

Performance-oriented motivations	Business-ecological motivations
Customer- and market-specific development	Sustained growth of Chinese economy and market size
Speed	Government policy
Supporting local production processes	Peer pressure
Cost advantages	
Input-oriented motivations	
Availability of qualified personnel	
Local pockets of innovation	
Tapping informal networks and knowledge sources	

Source: Author; Gassmann and Han (2004)

Performance-oriented motivations

Customer- and market-specific development

One main reason why so many companies are establishing development bases in China is to develop products locally specifically for the Chinese market. The necessity of adapting products to a foreign market is a widely shared belief among many interviewed R&D managers. Selling products without paying attention to the local market's needs is bound to fail. Locating R&D activities in China allows MNCs to adapt and tailor their products and services to the local culture, market needs and industry regulations. A typical example is adapting IT user interfaces, telecommunications or car infotainment products to be used with the Chinese language. Moreover, the specific local conditions in which products operate require appropriate modifications and redevelopment. For example, in China there is little customer demand for artificial tanning products, since light skin is more the ideal of beauty. Thus, skin- and body-care companies need to develop products which can make skin lighter. Geberit, a leading Swiss specialist in sanitary technology, had difficulties in selling its products at the beginning of its undertaking in China. The reason was simple: its products, designed in Europe, often did not match the Chinese construction layouts.

According to industry experts, over 75 per cent of the growth in electronic manufacturing capacity in the next two to three years will take place in China. The risks associated with shifting production for purely cost reasons are high; but local development and product adaptation in these growing markets can support manufacturing operations and increase competitiveness. There is an additional benefit to operating in cultures and countries such as China: a company can develop new products and forge advanced thinking on many product issues. Some managers believe that products which satisfy the requirements of the most difficult consumers and market environments are likely to succeed anywhere in the world. The Microsoft Research Centre in China is tackling problems associated with computing in Chinese due to the difficulty in inserting Chinese characters on a Western keyboard. Besides the improvement in software suggestion and error-checking systems, researchers also examined data entry methods such as speech and handwriting recognition. The result will make computers user-friendlier in Chinese, but will eventually benefit all computer users (Gelb, 2000). In the late 1990s the elevator and escalator company Schindler established an R&D centre in Shanghai due to its reputation as one of the most booming and sophisticated markets in the construction business. Chinese customers are less risk-averse than Western customers, which is typical of booming economies.

In 2003 Schindler conducted a field study for a new web-based personalised infotainment system in the elevator cabin – an advanced experiment that was less likely to be accepted in Europe or the USA. Based on that study, Schindler plans to multiply the system requirements for a product launch worldwide.

The manufacturing of products leads to product adaptation through R&D, which could in turn create product innovation. It is worth mentioning that many MNCs have entered the Chinese market as suppliers to long-term international business partners. As manufacturers have to adapt their products to the Chinese market and come up with new products regularly, the suppliers that have followed them are in the same situation. Most manufacturers have their own local product development teams, thus the supplier having a local R&D team guarantees smoother cooperation.

Cost advantage

In many industries, success increasingly relies on the company's innovativeness. At the same time, more innovation leads to higher R&D cost, which has in fact grown steadily over the decades. To date, the software industry spends 60–70 per cent of its total costs on R&D, the aviation industry 20–30 per cent and the automobile industry 10–20 per cent. Being productive at a competitive R&D cost level is therefore a strategic goal for many MNC chief technology officers. To run R&D facilities in China similar to those in the USA would cost only about one-tenth of the amount (Li and Zhong, 2003). The lower wage structure in China is one reason for this. Although the wages of highly qualified Chinese R&D professionals are high compared to the Chinese domestic level, they are only between one-third and one-fifth of R&D professionals' salaries in the triad regions. Companies in the software industry that moved to India are now starting to evaluate China, due primarily to the labour cost increases in India. Moreover, as mentioned above, China has enacted a series of preferential policies to encourage the establishment of foreign R&D activities, e.g. foreign R&D centres in China can import certain equipment duty-free, which provides further cost advantage potential for doing R&D in China. However, there has been a rapid cost increase, especially in salaries, in the top-tier cities such as Shanghai and Beijing. This also helps to explain the trend of more and more MNCs setting up their R&D centres in large second-tier cities such as Nanjing, Hangzhou, Shenyang and Xian.

Speed

With more demanding customers and increasing global competition, companies struggle to introduce products to the market quicker. Localised R&D allows for a shorter R&D cycle time, especially for products that require customer- and market-specific accommodations. Speed is one of the most important competitive advantages in a dynamic market such as China. For R&D centres that have been integrated into the MNC's global R&D network, the Chinese R&D site is an important cog in the machine to achieve a 24-hour development cycle. For pharmaceutical companies such as Novartis and Roche, the availability of large patient pools in all therapy areas and rapid, uncomplicated patient recruitment for clinical trials enable them to accelerate their research processes.

Supporting local production processes

Local R&D activities can assist manufacturing operations in improving quality, learning to produce new offerings, reducing costs and increasing capabilities and new products' speed of localisation. For example, the packaging specialist SIG opened an R&D unit in Suzhou in 2003 to support its manufacturing units.

Business-ecological motivations

Sustained growth of economy and market size

Despite the stagnated world economy, the national economy's dynamic growth and its overwhelming market size have made China one of the most strategically important markets for MNCs. Especially in IT and telecommunications fields, multinational giants such as Microsoft, Nokia, Motorola and Siemens have invested hundreds of millions of dollars in their R&D activities in China, which is in essence an investment in China's future market. For example, China has become the world's largest mobile phone market with more than 377 million subscribers at the end of September 2005 (China Information Industry Net, 2005). The critical mass of the Chinese and Asian markets is increasingly influencing mobile phone size, style and applications globally. As a response to the strong demand for affordable, entry-level mobile phones in the Chinese and Asian markets, the Siemens Mobile division in October 2002 established its global headquarters for voice-centric mobile phones in Beijing. According to Siemens, this marked the first time that the central responsibility for one of Siemens Mobile's business activities had been located in China.

Government policy

For over a decade, China's increasing sensitivity regarding technology's contribution to economic growth has been evident. 'Revitalizing the Nation through Science and Education', a strategy officially adopted in 1995 by the Chinese government to speed up scientific and technological progress (OECD, 2002b), has led to the rapid growth of China's national S&T activities. In 2000 China spent US$11 billion on R&D, which amounts to 1 per cent of its GDP. This represents an increase of 150 per cent over the 1995 level of US$4.2 billion (China Statistics Bureau, 2001). Furthermore, although OECD countries provide over 90 per cent of FDI globally, their share of FDI in China is much smaller. The major source of FDI in mainland China remains the Greater China regions such as Hong Kong and Taiwan. Therefore, China still has the potential to attract long-term, relatively capital-intensive and high-tech projects from multinational enterprises in OECD countries (OECD, 2003a). To realise this potential, China has continued to liberalise the approval process for FDI and a number of preferential policies have been put in place to encourage foreign businesses, especially MNCs, to set up local R&D investments. Chinese policy-makers believe that an effective way to bridge the international technology gap is to intensify the linkages with the foreign R&D community.

One important means of doing this chosen by the Chinese government was the establishment of high-tech parks combined with incentives such as free rent, low tenancy costs, favourable lease terms and tax relief (see Chapter 3). As identified by Ambrecht (2002), there are several multifaceted reasons behind these kinds of policies. Foreign laboratories will bring capital investment, ancillary spending and job opportunities. They help to attract excellent ethnic Chinese specialists from around the world back to China to conduct advanced research. The proximity to international research facilities will also spur the Chinese higher education system through their demand for local high-quality technical personnel and cooperation with Chinese R&D facilities. Moreover, the business background of these R&D laboratories could help China to create market value from the leading-edge technologies being developed in Chinese universities and research institutions.

Local R&D activities are considered to be important evidence that a company is interested in developing long-term commitments in China. It helps to build trust and good working relations with the government and to obtain official support. However, due to financial incentives and other business advantages, some foreign firms are tempted to register their China activities as 'R&D' no matter whether or not their work entails genuine research and development activities. Chinese government officials may be more inclined to approve business licences for other MNC activities if 'face' is given via the establishment of a local R&D centre (see Walsh, 2003; Han et al., 2005).

Given the pure power of the Chinese government, it is in a position to pursue the so-called 'technology for market' policy in order to accelerate MNCs' investment level and R&D commitment. Prior to China's accession to the WTO, foreign investors were regularly pressurised to transfer technology in return for market access. For example, there was intense competition between several global automobile companies concerning the establishment of a joint venture in Shanghai in the late 1990s, which was speculated to be the last approval of this kind for many years. General Motors finally won the licence to establish a joint R&D centre (*The Economist*, 1999; Walsh, 2003). In this particular case, political pressure from the Chinese government had primarily determined the MNC's R&D investment decision.

Peer pressure

MNCs' motivations for establishing local R&D in China are rooted in awareness of the possible mid- and long-term competitive advantages discussed above. Given the increased competition among MNCs in the Chinese market, those which do not have foreign R&D centres have come under increasing pressure to invest in R&D. Even though most interview partners did not want to admit peer pressure as a driver, it was mentioned during informal follow-ups.

Input-oriented motivations

Availability of qualified personnel

Among the input-oriented motivations for establishing R&D in China, the huge human resource potential is of great importance. In 2002 2.5 million students graduated from the country's 2,000 higher education institutions (HEIs) and colleges, including more than 66,000 masters' graduates and 14,000 PhDs, ranking China third behind the USA (around 40,000 PhDs) and Germany (around 30,000 PhDs) (Ministry of Education, 2003a). Between 1998 and 2004 the number of graduate students quadrupled to over 270,000. The total on-campus student population exceeded 20 million in 2005 (*People's Daily*, 2005b). By 2010 the Chinese government aims to provide access to higher education to 23 per cent of the college-age population (*People's Daily*, 2003a). Many top-flight universities such as Tsinghua, Peking, Zhejiang and Fudan produce highly qualified graduates in disciplines such as mathematics and natural sciences.

Aside from domestic graduates, more than 580,000 Chinese students travelled to 100 countries and regions worldwide for studying purposes between 1978 and 2002, with 150,000 of them returning to China (Ministry of Education, 2003b). The majority of students chose to immigrate to other countries after graduating. Hence, China has to a great extent suffered from an outflow of talent (brain drain). In recent years Chinese governments, at both national and local levels, have introduced policies to induce highly skilled overseas Chinese to return to China (OECD, 2002b). Increasing numbers of scientists and graduates have returned from abroad thanks to the enduring economic growth and improved opportunities in China. In 2002 more than 18,000 'returnees', as they are called, came back to China, an increase of 47 per cent compared to 2001 (Ministry of Education, 2003b). These key people represent an additional pool of uniquely qualified human resources, and bring experience and knowledge from around the globe.

Tapping informal networks and information sources

In China, business success is heavily dependent on strong informal networks and relationships – the frequently cited *guan xi*. The establishment of a local R&D centre enables a company to build and maintain informal networks with universities and local scientific communities, which can help MNCs to establish strategic partnerships and secure high-quality human resources in the long term. In addition, China's industrial development is at an emerging level and the economy is undergoing a transition from a planned to a market-based system. Hence, changes in industrial regulations, legislation and policies are more dynamic. Their on-site R&D activities and proximity to the government, industry associations and standard-setting authorities help

MNCs to keep pace with and even influence changes in this dynamic environment, while also allowing them to achieve critical competitive advantages. Using local R&D to gather technology intelligence (listening posts) on local and international competition is yet another input-oriented motivation for MNCs to conduct R&D activities in China.

Local pockets of innovation

Since Chinese policy-makers seek to raise the level of China's industrial production and increase competitiveness to an international level, special economic and other investment zones have been established and have become the main engine for growth in the Chinese economy. Notably, high-technology development zones (HTDZs) or 'science parks' have been designed to lure researchers, entrepreneurs, foreign R&D centres and venture capitalists from around the world. An example is Beijing's high-tech Zhong Guan Cun area, located north-west of the city and home to some 30 universities, including leading institutions like Tsinghua and Peking Universities, and more than 200 research bodies, including the Chinese Academy of Sciences as well as a number of spin-off enterprises from these universities and research institutes. As a result, there are a number of start-up firms, foreign-capital firms and large-scale local firms seeking access to high potential by building strong relationships with universities. Furthermore, these pockets of innovation attract investors by providing space, advanced infrastructure and the high-tech facilities they need, besides financial incentives. The Chinese State Council and Beijing municipality, for example, both offer start-up firms located in the Zhong Guan Cun area tax-free operation for three years after establishment, followed by a 50 per cent reduction in the next three years and a 15 per cent discount from the seventh year onwards, along with other tax incentives. Due to the substantial governmental support and geographical uniqueness, it is not surprising that several interview partners believe these industrial and science parks will become centres of excellence in the future.

Challenges of managing R&D in China

China is a very attractive location for MNCs' R&D units. But despite the advantages and rewards, there are still high managerial barriers that could neutralise the potential advantages. This section will describe the managerial challenges that international R&D centres face in China. As stated by Perrow (1970), barriers to effective managerial practices are derived from two situational characteristics: complexity and unpredictability. According to Smith et al. (1996), complexity occurs when problems that managers encounter are multifaceted and methods for dealing with them are diverse. Unpredictability impacts on the flow of organisational processes and managerial actions and causes uncertainty in managing a company. Most of the identified barriers to managing R&D in China result from both complexity and unpredictability, and cannot be attributed to a single characteristic.

The identified barriers of managing R&D in China have been divided into two categories: challenges at the intra-organisational level and challenges at the inter-organisational level (see Table 5.2). Challenges in areas such as human resources and intellectual property were mentioned throughout each interview, whereas sensitive topics such as governmental influences and bureaucracy were only touched on by a few of the interviewed managers.

Table 5.2 Identified challenges of managing R&D in China

Intra-organisational challenges	Inter-organisational challenges
High employee turnover rate Low individual initiative and innovative mindset Difficulties in managing Chinese R&D professionals due to cultural differences and team diversity Remaining governmental influence within joint ventures	Infringements of intellectual property rights Bureaucracy Uncertainty regarding cooperation Uncertainty regarding regulation changes

Challenges at the intra-organisational level

High employee turnover rate

Like other foreign enterprises in China, many R&D departments are plagued by high staff turnover rates, particularly those located in large cities such as Beijing and Shanghai where plentiful new opportunities are available. In general there are three main causes of staff turnover: some employees leave because they have simply found a better-paid job elsewhere; some go abroad to obtain graduate or postgraduate degrees; and a few leave to work for, or establish, high-tech start-up enterprises (Walsh, 2003). This is a common phenomenon that affects not only foreign companies in China, but Chinese domestic high-tech companies and research institutes as well. Foreign companies are often used as a career springboard. Working for a foreign company not only provides Chinese graduates with a higher salary and practical experience, but also familiarises them with Western management practices and provides possibilities of advanced on-the-job training.

References from these jobs will help them get jobs with better pay and possibilities. The worst-case scenario is that they use this experience to be hired by competitors. As stated by several foreign managers and confirmed by Chinese employees, compensation strongly influences the affiliation and loyalty of Chinese R&D staff. As a result, some R&D managers expressed a reduced trust in local staff, since they are afraid that this lack of loyalty will lead to a brain drain to competitors. This is especially relevant within Sino-foreign joint ventures where the Chinese partner's parent company, or simply the partner itself, participates in or owns other domestic enterprises operating in similar business fields. For example, ABB, a power and automation technologies giant, has lost much technological knowledge through its Swiss-Chinese joint ventures. Building long-term staff loyalty is a challenge for human resource managers in China. It is particularly relevant for R&D labs, given that know-how travels with people.

Low individual initiative and innovative mindset

The majority of local R&D staff are recruited from China's leading universities. During the interviews, most managers shared the opinion that Chinese graduates have a solid education and are highly skilled in solving certain well-defined tasks. But there is an awareness of a lack of practical experience and individual initiative, which is, to a certain degree, in line with Walsh's (2003) findings. It could be argued that this phenomenon is attributable to the Chinese education system, which is characterised by a narrow curriculum design and very little development of individual initiative. To a great extent, an R&D staff member's individual initiative is decisive for creativity and innovation. As Walsh (2003) stated, developing a more innovative mindset among Chinese staff is a primary concern for foreign R&D managers at this stage. Risk-taking behaviour and entrepreneurship in the widest sense have to be promoted. As a result, the management and development of R&D staff in China require much attention.

Difficulties in managing Chinese R&D staff due to cultural differences and team diversity

Given the general lack of experienced indigenous R&D managers in China, many top R&D managements are staffed by expatriates. Unfortunately, most of them do not have adequate experience in the Chinese environment. The Chinese language is no longer much of a barrier, as most Chinese researchers have a good command of English. A larger obstacle for Western managers is to overcome cultural differences during daily interactions concerning issues like communication style and 'face-saving', to name just two. A Western manager may have done everything correctly according to his understanding of good management style and 'face-saving'. However, lack of experience and sensitivity to the Chinese mentality and culture will usually translate into managerial inefficiency, wrong decisions and inadequate leadership. Western managers, who mostly come from 'low-context' cultures (e.g. Germany, the USA), are used to capturing a message's meaning with words alone. They believe spelling it out clearly is the only way to avoid ambiguity. Conversely, the Chinese culture is very 'high context' (Hofstede, 1994): a message is delivered with non-verbal signals (e.g. tone of voice, use of silence, facial expressions and body language) and unspoken assumptions in the context or environment surrounding the conversation. People from high-context cultures assume that the receiver of the message is intelligent enough to understand its true connotation. A lack of awareness and proper handling of the barriers between high- and low-context communication styles can eventually lead to misunderstanding, confusion and inefficiency.

The R&D teams of MNCs in China are diversified, and typically composed of three groups of people: local graduates make up the majority of the staff; Western expatriates and 'global' Chinese form the remainder. Although diversity in R&D teams can increase creativity and innovation, it can also be a source of potential conflict. In addition to the general difficulties of managing intercultural teams, one particular challenge, which is often overlooked, lies in the potential differences between local Chinese staff and the global Chinese. In this context, 'global Chinese' is a generic term for three subgroups of Chinese people working for foreign R&D: mainland Chinese returnees with study and work experience abroad; Chinese from Greater China (i.e. Taiwan, Macao and Hong Kong); and overseas-born ethnic Chinese. On the one hand they share the same Chinese origin and culture and have almost no language difficulties with one another. On the other hand, due to multilayered differences between them stemming from such elements as different educational backgrounds, working styles and perceptions, and in particular the huge gaps in pay for the various levels (De Boer et al., 1998), one should be wary of generalising these three subgroups. Western expatriates are often not aware of or underestimate these differences.

Remaining governmental influence within joint ventures

It is worth mentioning that government influence within joint ventures is a further management barrier in China.

Even if intervention in foreign enterprises' activities by the Communist Party of China (CPC) has decreased in recent years and the party branch (i.e. party secretary) within some wholly foreign-owned companies is not involved in the business at all, some interviewees still mentioned the strong governmental influence. As one manager said: 'There are still numerous possibilities for the Chinese government to make everything difficult.'

Challenges at the inter-organisational level

Intellectual property rights

Given the challenging nature of the Chinese economy, foreign investors must assume some inadvertent technology transfers. As Schumpeter (1950) noted, innovations should lead to temporary monopolistic profits in order to harvest previous R&D investments. Therefore, a strict legal intellectual property system with little uncertainty for the innovator is a prerequisite for technological progress and high rates of innovation.

The current Chinese intellectual property rights (IPRs) framework has been in existence for less than two decades (Yang, 2003a). China has made progress in protecting IPRs and has ratified different international treaties and conventions to show the world that it strives to be in step with international intellectual property (IP) standards, yet IP piracy remains rampant in China. In-depth interviews revealed that IP is an area about which foreign enterprises have serious concerns. The International Intellectual Property Alliance estimates that over 90 per cent of business computer software in China is pirated (IIPA, 2003). According to estimations, Microsoft loses several billion dollars due to piracy each year. One major reason for this kind of concern is weak IP adherence enforcement. Starting and winning a patent case in China are still difficult, and definitely time- and energy-consuming. Additional concerns for foreign companies regarding IPRs include lengthy patent application procedures and a lack of public acceptance of IPR legislation (OECD, 2003b). Sometimes the lack of IP protection stops foreign firms from transferring their cutting-edge technologies, research or equipment to China.

Nevertheless, the number of foreign applications for patents attributable to high-tech industries has continued to rise. This is mainly due to increased competition and a governmental policy that transfers technology to China. IP management in China is intricate, but achievable if enough effort and resources are put into it. For example, Siemens' corporate research lab in Beijing has a special group dedicated to IP issues in China. New technologies stemming from wholly foreign-owned R&D centres in China are initially filed, and in some cases exclusively, in the foreign company's home country. In general, when conducting innovative research in China at present, foreign investors are very sceptical of IP protection, consequently truly innovative results in China will still take some time.

Bureaucracy

As mentioned above, the Chinese government provides incentives for foreign R&D activities in China. According to several of the interviewed R&D managers' experiences, actually receiving the promised preferential conditions such as tax relief and other incentives can be a stressful and prolonged procedure, due to multiple bureaucratic hurdles and very specific rules. Importing test materials can be difficult, and transferring people from Beijing to Shanghai requires an official permit. Therefore, a good relationship (*guan xi*) network with the government is crucial for business efficiency and success.[1] This kind of relationship network needs time and occasional donations. As one Western expatriate manager mentioned, relationship investments with local governments take the form of sponsoring IT equipment for local universities, or other contributions to non-profit official organisations such as a municipal kindergartens. However, one should not mistake this kind of financial aid for a bribe.

Uncertainty regarding cooperation

Industry-university cooperation (IUC) is well known for its potential to yield mutual benefits. Especially in R&D, IUC could offer strategic opportunities and increase productivity regarding innovation. In contrast to the universities of developed countries, Chinese universities and academics still do not have sufficient experience in cooperating with industry partners, as IUC is still not broadly institutionalised in the country. Over the years several big MNCs, such as Microsoft, Lucent Technologies and Nokia, have accumulated substantial experience in working with Chinese universities and research institutes. However, many international R&D centres, especially the new ones, do not feel confident when cooperating with a local university. Thus, for many foreign R&D centres, the huge potential of industry and university cooperation in R&D remains unexploited. Furthermore, as was often mentioned by interviewed managers, there are uncertainties concerning the fairness of R&D cooperation, due to insufficient legislation in this regard (especially IPRs) and the regional governments' strong protectionism. Many local cooperation partners are state-owned enterprises (SOEs) and spin-offs from former SOEs. Unwanted and non-transparent knowledge/technology transfers can be carried out within this network – something that is not obvious to an outsider. Even if the foreign partner has clear evidence of IP infringement, its means of counteracting it are limited.

Uncertainty regarding regulation changes

Due to Chinese policy-making's lack of transparency, foreign parties find China's industrial, political, legal and technological policies and strategies difficult to discern. This makes foreign R&D activities in China even more uncertain. For instance, in the automobile industry China sees Europe as a role model with regard to industry norms and regulations. As a result, the Chinese government ambitiously follows the European automotive norms (e.g. emission norms equal to EURO III and IV, airbag obligation, tyre pressure control and brake power control). The validity of these norms in respect of automotive products is thus left to the government's discretion. This kind of unpredictability makes R&D activities even more challenging.

Conclusion of research phase I

This chapter has discussed the findings of the first research phase. From three different perspectives, i.e. performance-oriented, business-ecology-oriented and input-oriented motivations, the author has built a cluster of various determinants for MNCs' R&D investments in China. As the in-depth interviews revealed, the launch of the majority of foreign R&D centres under investigation was initially driven by performance-oriented motivations. The fast-growing Chinese market is a further major incentive for foreign R&D

to enter China. The goals of achieving customer- and market-oriented development for a market with tremendous potential, and with the help of qualified personnel who are mostly available, are the two most important incentives for foreign R&D to enter China. It is worth mentioning that performance-oriented motivations are much more in line with the home-base-exploiting point of view, in which R&D facilities are associated with the adaptation and modification of existing technological assets in response to local conditions. In other words, local engineering and R&D capabilities are installed to support local production facilities and adapt products to Chinese needs.

The interview results also suggest that an input-oriented motivation has increasingly started to emerge in China, which is consistent with the home-base-augmenting theory. Within the context under investigation, input-oriented motivation not only refers to specific local knowledge, e.g. very detailed knowledge on TCM in the Unilever case, which is not available elsewhere in the world, but also to valuable local talent, such as very important corporate assets. As stated by several interviewees, one of the most important reasons for MNC R&D investments is that these MNCs have recognised the high potential of the country's human resources. However, human resource management is also frequently mentioned as one of the major challenges in managing R&D in China. To date, home-base-augmenting and home-base-exploiting R&D exist in parallel, although the latter still forms the majority. As the country's scientific maturity is increasing steadily, it will not be surprising if more HBA R&D activities appear in China in the near future.

The HBA and HBE theory does not pay attention to business-ecological determinants such as governmental interference and relationship-building aspects. The Chinese government still plays a decisive role, especially in the economy. For Chinese leaders the strategy for economic growth is quite clear: science and technology are fundamentals for catching up with the world's leading economies. As a result, the government has introduced different political and regulatory means to accelerate the establishment of foreign technology in the country, although the immature IP protection system remains the most serious obstacle to achieving this aim. Sometimes the government has also pursued what is termed a 'technology for market' policy; in other words, in certain industries only MNCs' state-of-the-art technologies could gain market access. However, in China the market often outweighs the government, and since competition is becoming fiercer, the market rather than governmental regulations forces many MNCs to bring the latest technology into the country.

Doing R&D in China is a strategically important and also logical move for many MNCs. However, to manage R&D activities in China successfully, managers still have to overcome several important barriers. The interviewees frequently underlined three of these: managing R&D professionals, IP and industry-university cooperation. In research phase II the author attempts to develop a more thorough understanding of the challenges and uncertainties that foreign R&D managers face in these areas, and discloses and develops important managerial implications.

Notes

1 This might not always be the case. One of the interviewed high-tech companies has applied to the central authority to set up a corporate post-doctoral station. Somehow, even with the support of the local government, there has been no progress.

6

Managing R&D professionals in China

Among the critical challenges faced by executives of foreign R&D activities in China, the management of R&D professionals is one of those most frequently mentioned. To a certain degree this is not surprising, since human resources (HR) are essential and often the most important element for a successful operation. Within the context being researched, three factors make human resource management (HRM) even more critical. First, an R&D activity is *per se* strongly knowledge-based and knowledge travels with people, especially tacit knowledge. Hence, proper HRM ensures high-quality R&D achievement. Second, the Chinese environment as such is extremely dynamic. A successful operation within a dynamic setting strongly depends on the continuity of dedicated resources, in particular HR. Third, the Chinese culture is alien to most Western R&D executives. The possible gains that foreign R&D in China might bring could be outbalanced by risks such as a cultural clash in managing local R&D professionals. Without a thorough understanding of HRM practice and local culture in China, MNC R&D activities will not be able to take full advantage of the available opportunities. Based on the findings of research phase I, this chapter aims to unlock HRM practice and its managerial implications in managing

R&D professionals in China. The following sections first review the existing knowledge in terms of HRM within the Chinese context and outline the primary focuses and findings of previous research. Particular attention is also paid to previous literature on HRM issues in R&D management. The research questions will then be posed. Prior to revealing the research findings, it is instrumental that the new generation of R&D professionals in China should be understood. Finally, managerial implications drawn from the field research are discussed.

Strengths, deficits and particularities of R&D professionals

According to research by von Zedtwitz (2004), more than two-thirds of MNCs' initial R&D directors in China were non-Chinese expatriates, and most had little or no experience of China before being assigned to the job. Although these expatriates might have received preparation courses for their function within this new working environment, the Chinese context remained alien to most. Many aspects, such as working habits, leadership style and employee motivation, differed from what they were used to. It is, without doubt, a big challenge to manage and lead a group of R&D professionals, or inexperienced graduates, from a totally different cultural background and get them to perform. A building-block that is instrumental to the successful practice of HR in recruiting, staff training and development is an in-depth understanding of local R&D professionals, involving thorough learning about their strengths, deficits and particularities. Based on this insight, differences and obstacles can be bridged by adapting to each other's culture. This section outlines the characteristics identified with regard to local R&D professionals' and graduates' strengths, deficits and particularities, and their basis.

As mentioned earlier, foreign R&D centres in China are a relatively new phenomenon. Between 2003 and 2005 the number of such centres rose from around 200 to more than 700. This dramatic boost created a higher demand for R&D professionals than the HR market in China has ever encountered. However, during the first three decades of the PRC, Chinese policy-makers failed to build up an advanced education system. As pointed out by Fischer and von Zedtwitz (2004: 354), 'China explicitly chose to emphasise primary education and basic literacy at the expense of more advanced education.' The Cultural Revolution (1966–1976), during which most secondary schools and universities were closed, was responsible for a 'lost generation' in terms of education. It explains why there is a lack of experienced scientists and engineers in China aged in their mid-40s and mid-50s, who ought to be filling senior positions in the S&T area. This phenomenon also applies to university faculties and management positions.[1]

The identified strengths, deficits and particularities with regard to local R&D staff should not be considered as a mere stereotyping of Chinese scientists and R&D professionals. They are the shared cognitions and conclusions of the interviewed managers and R&D executives, based on their experiences and observations. The majority of international companies' R&D staff in China are fresh graduates from the country's leading universities; thus the characteristics of Chinese R&D professionals also strongly reflect Chinese graduates' characteristics. At the same time, this observation reflects a society's transition, its young people[2] and their latent potential. It is worth mentioning that among the interviewed executives there are overseas Chinese who have received an international education and gained work experience abroad. Their perceptions could provide more objectivity, since they are familiar with both 'worlds' and are able to develop a more grounded understanding.

Strengths

Table 6.1 lists the identified strengths of Chinese R&D professionals and graduates as acknowledged by the interviewees. They are consolidated into two different categories: technical strengths and exemplary attitude.

Technical strengths

Chinese R&D professionals and graduates are described as having an *excellent basic education* and *solid technical knowledge*. One obvious reason for this judgement is that most graduates hired by MNC R&D centres were top of their class at the country's leading universities and offer outstanding qualifications.[3] The Chinese education system and methods contribute to these extensive strengths. Learning at Chinese schools and universities is still influenced by the country's Confucian heritage, which is rote-based. Students spend most of their time memorising

Table 6.1 Acknowledged strengths of Chinese R&D professionals

Technical strengths	Exemplary attitude
Excellent fundamental education	Ambitious and highly motivated
Solid technical knowledge	Diligence and readiness to perform
	High willingness and ability to learn

and repeating assigned tasks and contents. Although Western countries consider this obsolete from a pedagogical point of view, it leads to Chinese graduates having an excellent command of the basic technical skills in mathematics and the natural sciences. On the other hand, this kind of education has strong drawbacks, such as a lack of the innovative and critical thinking that is essential for R&D. This issue will be discussed in more depth later in this chapter.

Exemplary attitude

Throughout the interviews with executives and R&D managers, Chinese R&D professionals and graduates were frequently characterised as *highly motivated and ambitious*, *diligent* and *highly willing to perform*. Moreover, many foreign R&D managers are impressed by their *ability* and *eagerness to learn*. Besides individual factors, a logical reason for the outstanding motivation displayed by Chinese R&D professionals and graduates is China's increasingly competition-oriented modern society. From primary school onwards, most students' activities are directed at the 'examination of life', i.e. the university entrance examination that decides whether one will belong to the white- or blue-collar world in future. Starting very early, most Chinese children are taught to excel in terms of school performance. Parents, especially those who could not go to school or university during the Cultural Revolution, hope that their children will have access to higher education. A child's education is usually the most costly, and thus most important, investment that an average Chinese family makes. In combination with the 'one-child' effect, young people experience extremely strong pressure to succeed. Although the enrolment criteria for higher education have been widened since 1999 (Ministry of Education, 2004), competition is fiercer, simply because the number of applicants is also increasing rapidly. The country as a whole believes that education is the only way to wealth.

Interviews with young Chinese R&D professionals reveal that working for a leading international company means self-realisation for many Chinese graduates after years of never-ending studies. Such companies appreciate their work, since they know the quality of those with whom the graduates had to compete. Several interviewed managers mentioned that Chinese R&D professionals are more willing to work overtime than their foreign colleagues. They also do more than is asked, not only because of the higher pay and social standing that MNCs provide, but also for the unique learning opportunities and valuable experience offered by MNCs, all of which will ensure their future career in China. As one R&D manager underlined, 'They want to be led just as their colleagues are led in the West.' They are aware that by working hard they can achieve things that their parents never had the chance to achieve. Keeping this in mind, many Chinese R&D professionals' exemplary, highly motivated attitude towards work and willingness to perform are almost self-explanatory.

Deficits and particularities[4]

The findings discussed in the last section do not seem very surprising, since Chinese people are well known for 'working hard'. As indicated by the field research, young Chinese R&D professionals and graduates also reveal several typical deficits and particularities related to their work. The identified issues are divided into two categories: deficits and particularities regarding R&D-related skills, and those regarding complementary skills (see Table 6.2). To a large degree, weaknesses frequently mentioned by young Chinese R&D professionals and graduates reflect the Chinese education system's deficits. Nor should the impact of culture be ignored.

Deficits and particularities regarding R&D-related skills

As stated by one interviewee from the high-tech industry, it could take up to three years of work experience for young Chinese R&D professionals to catch up with their US counterparts in terms of productivity. This huge productivity gap has different causes. *Lack of innovative thinking* is one of the serious deficits many MNC R&D managers find in their local staff. The question that should be asked here is why this is so. Several aspects of an educational, cultural and political nature impede the innovative and *creative thinking* ability in China. Didactics at Chinese schools are very much teacher-centred and lack reflective learning. Usually learning is based on the use of examples and existing models, and students are expected to absorb and copy what

Table 6.2 Acknowledged deficits and particularities of Chinese R&D professionals

Deficits and particularities in R&D-relevant skills	Deficits and particularities in complementary skills
Lack of	*Lack of*
Innovativeness ■ creativity ■ critical analysis ■ networked thinking	Communication ■ verbal English language proficiency ■ presentation skills ■ delivery of feedback
Ability to work independently	Teamwork ability
Structured problem-solving skills	
Hands-on capability	

is taught (see also Chapter 7). Chinese teachers often do not have the capacity to encourage their students to think individually when facing a class of up to 50 pupils or students. Besides, teachers too lack the ability to promote creativity and *critical analysis*, as they have been taught in the same way. The situation does not differ much at higher education institutions. Professors often simply read from the textbooks. To a certain degree, this kind of education practice is rooted in Confucian thinking, which is still the foundation of modern Chinese society. As put by Liu (2003), 'Geert Hofstede attributed "Confucian Dynamism" to "a society's search for virtue" rather than a search for truth' (see also Hofstede, 2003).[5] This is also a reason why China lacked a solid scientific foundation despite its rich civilisation. Moreover, from a political point of view, more than 56 years of a communist-socialist system certainly did not encourage the questioning of authorities, nor were superiors' ideas often challenged. These aspects led to a long tradition of Chinese people not being encouraged to be creative or critical of their superiors. This explains young Chinese students' deficits regarding 'breaking the rules' and thinking innovatively. As pointed out by one of the interviewees, it changes the way problems are addressed. Occasionally the lack of critical thinking can be traced to a kind of mental barrier or lack of *self-confidence*. It can even be rooted in the attitude that 'all things that come from the West are good', which is prevalent after the country's long isolation.

The limited freedom for self-determination, no matter whether at school, at university or at home, also affects the ability of Chinese graduates to *use their own initiative* and *work proactively and independently*. Young people in China have been raised in very protective families. Due to their extremely sheltered childhood, youth and student life, many young people, especially those from large cities, have not had much chance to explore things independently. They are less socially mature than their Western counterparts, who often achieve independence much earlier. The interviewed managers also mention that they often encounter an *unstructured approach to problem-solving* due to their young employees' weak ability to think independently. Other comments were made by many interviewees: 'Chinese R&D staff perform very well, once a task and the way to the end result have been clearly defined'; 'their *networked, combinatory and holistic thinking* is weak, as professors often let them do very narrowly defined tasks'; and 'once an instruction is given, they move straight forward without looking around'. However, this is a poor working style for doing R&D, as the 'ways' to the desired results are often unknown. Depending on the type of R&D activity (research-oriented or development-oriented), this kind of deficit could be disadvantageous to a greater or lesser degree.[6]

In addition, as in many other countries, executives complain that graduates leave university with too much theory and too few practical skills needed in the workplace. According to a software company executive, Chinese universities' IT faculties do not have enough resources to train students in IT hardware, due to the high cost involved, hence Chinese IT graduates are better at software programming than at dealing with hardware. This is also a reason why many foreign high-tech companies start their R&D activities in China by concentrating on software development first. Moreover, Chinese graduates' knowledge is often characterised as too 'vertical' (more depth) and less 'horizontal' (less breadth), which is quite the opposite of the Western concept in which a broad knowledge and solution-oriented thinking are advocated in education. For example, in the software industry, knowledge about the market and the customer is important. One interviewee noted: 'Chinese software engineers often have excellent technical know-how, but lack business know-how. To develop software that meets the market's needs remains a big challenge for them.'

Deficits and particularities regarding complementary skills

Chinese R&D professionals' communication skills are among their frequently mentioned deficits in complementary skills. This will be examined from two different perspectives: deficits in skill development *per se*, and differences due to cultural influences.

Young people's English language proficiency is improving rapidly. In fact, English is now considered one of the most important subjects at Chinese schools. Nonetheless, a typical phenomenon is that most Chinese students can read and write English at a solid level, but their *oral communication skills in English* are relatively weak. Furthermore, many young people are weak at presenting and speaking in public. Recent research shows that Chinese students' greatest fear is of speaking or performing in public (Davey and Fei, 2005). The lack of training opportunities in these skills at Chinese schools and universities is mainly responsible for their underdevelopment.

Chinese culture also plays its part in this respect. Managers complain that Chinese R&D staff are reluctant to speak during team meetings with foreign colleagues, even if they do have a good command of verbal English. An interviewed executive stated that during telephone conferences with American colleagues, the American side often talk most of the time and the Chinese side make only a limited contribution. Regardless of the question of individual personality, strong cultural influences underpin this situation, one of which is the 'face-saving' that is among the most distinctive characteristics of Chinese culture. Chinese people fear losing 'face', which means, for example, saying the wrong things in public or being criticised by others. And, vice versa, they rarely criticise publicly so as not to harm the 'face' of others, especially supervisors or people with seniority.[7] Furthermore, the preservation of group harmony is highly appreciated in this culture. Foreign executives of Sino-foreign JVs report that meetings are often just a formality to announce decisions rather than to make them based on open discussions. The real dialogue and decision-making takes place in private before or after the meeting. In addition, the 'high context communication style' (Hofstede, 1994) sometimes makes communication difficult through its imprecise expression.

Due to the reluctance of Chinese R&D staff to *provide clear and direct feedback*, foreign managers sometimes find it very difficult to estimate how well messages are understood and whether there are any questions. A typical situation many foreign executives have encountered is that after a briefing, everybody seems to understand what is required and the executive or other project partners no longer receive feedback from their Chinese colleagues. If this happens within a Western context, it means that everything is progressing smoothly and according to schedule. Yet within the Chinese cultural context, it is often different. After a while, executives sometimes realise that their staff have not understood everything exactly and progress has been faltering without being noticed. This kind of behaviour could endanger the project's success, since a development cycle is extremely short these days. The fear of 'losing face' makes Chinese people reluctant to provide direct feedback, or admit that they have not understood. Thus reluctance to *provide feedback* for cultural reasons needs to be taken into consideration in managing and communicating with Chinese staff.

Some managers realise that in China knowledge-sharing with colleagues is limited. This phenomenon is not unusual *per se*, as knowledge is power and people do not like to lose power. But to interpret this phenomenon correctly within the Chinese context, one should also include another dimension. The collectivistic Chinese culture is characterised by its strong differentiation between in-group and out-group members. Compared with individualistic cultures, the Chinese are more cooperative and supportive towards in-group members; they are more distrustful of and show stronger dissociation regarding out-group members (Triandis, 1995). Knowledge exchange occurs far more intensely within the in-group. This behavioural pattern also explains why Chinese team leaders and team members only care about their own tasks, and pay much less attention to other teams than their foreign colleagues do. The key to winning the trust of Chinese staff is to become an in-group member, but, simultaneously, not to forfeit one's authority.

Managing *teamwork* could be a further challenge for managers to face in China. As pointed out by several interviewees, many young Chinese graduates have difficulties with and feel uncomfortable working in teams, as they are not familiar with this type of work. Due to the 'one-child' policy, the majority of young people in China have not grown up with siblings. It often creates what is called the '4-2-1' dilemma, with four grandparents and two parents devoted to one single child. This constellation has consequences for the child's future development. A study of single-child families in China found that these 'little emperors and empresses' were more egocentric, less persistent and less cooperative than children with siblings (Jiao et al., 1986).

Managerial implications

The discussions in the last two sections have shown that the potential of young R&D professionals in China is high. They have distinctive strengths, such as an excellent basic education, *solid technical knowledge*, extremely high motivation and willingness to learn and perform. However, these strengths can be outbalanced by deficits such as lack of innovativeness and critical thinking, weaknesses in working independently and lack of hands-on capability. The field research also exposed deficits regarding complementary skills in areas such as verbal English, presentation skills and teamwork due to a lack of opportunity to practise them. As discussed previously, many of the 'deficits and weaknesses' are affected by the country's specific cultural, historical and institutional peculiarities. However, people can overcome these differences and achieve further development once proper circumstances and opportunities are provided. A good example of this is that in the era before the 'open-door' policy there were almost no entrepreneurs in mainland China, although the Chinese abroad have always been known for their entrepreneurial spirit and successes, whether in America or in South-East Asia. As pointed out by Fischer and von Zedtwitz (2004), it is certainly not true that 'the entrepreneurs have all emigrated, and those left behind are the ones without risk-taking genes'. The real reason is that the people who went abroad were given the chance. The given circumstances, combined with certain virtues, are responsible for the success of these overseas Chinese. The conclusion one can draw from this example is that circumstances can shape people and enable them to develop skills. The skill deficits of Chinese R&D professionals can be compensated to a great degree by providing proper on-job and off-job training and mentoring.

As Walsh (2003) stated, many Western managers want to change Chinese working habits and attitudes to conform better to Western business models and styles, while Chinese staff also seem keen to learn these methods and change their performance accordingly. Yet foreign managers should keep in mind that it will remain critical to manage particularities with deep cultural roots. Although some of these cultural attributes and values are being modified in accordance with the changing Chinese society (Jossi, 2000), one cannot fundamentally change culturally rooted behaviour by means of training and exercise. Culture as the 'software of the mind' (Hofstede, 1991) cannot be easily overwritten. That is why certain proven Western management styles might not succeed in China.

In order to deal with these differences or 'deficits', foreign managers should develop awareness of the differences and the underlying reasons. Proper adjustment to management practices that are based on an understanding of cultural differences can achieve positive results and cross-cultural advantages. One example illustrates how an interviewed MNC has made several adjustments to its project management practice in China by paying attention to the acknowledged differences. Officially, the working language at this company, as at many R&D centres, is English; in practice, English is only spoken between Chinese and non-Chinese colleagues due to the increasing localisation of many MNCs. This is one reason why the company tries to avoid Chinese-only project teams. A foreign team member acts as a project culture carrier, and also offers Chinese R&D staff a good opportunity to practise their verbal English. Moreover, efficient teamwork can be encouraged by experienced non-Chinese team members. In order to control the project's progress better, project teams in China meet more regularly. The team leader continuously reiterates the project's current emphasis and priority, and takes proactive control. At the same time, the size of the meetings will be kept small in order to create a familiar atmosphere for open discussions and encourage critical and creative thinking. However, this kind of setting requires the project manager's intense dedication and, ideally, extensive working experience in the Chinese environment. Overseas Chinese and 'returnees' could be the right candidates to take on these jobs, as they are more familiar with the context and the people than most of their Western colleagues (this issue is discussed in more depth later on).

This case depicts the possibilities to manage effectively some obstacles that foreign R&D managers could encounter in China. It is not a universal panacea, but gives an example of how deficits and differences could be overcome through adjustments to management practice. The conclusion should be drawn here that a deep understanding of the context is the most important prerequisite to enable the interaction of culturally diverse staff and achieve positive outcomes for both parties. Despite the possible frustrations foreign R&D managers may encounter, they should keep in mind that young Chinese R&D professionals are able to exert much energy if they receive the right training and are properly managed.

Recruiting

Just as the number of R&D professionals hired by foreign R&D firms is increasing steadily,[8] so is the huge oversupply of university graduates as potential candidates for R&D

jobs. In the past few years China has increased the percentage of high-school graduates who have access to higher education institutions (HEIs) from 4 per cent to a gross enrolment rate of 19 per cent. The total on-campus student population exceeded 20 million at over 2,000 HEIs in 2005 (Ministry of Education, 2005). This has resulted in an explosion of university graduates; in 2002 there were 1.5 million, but they were projected to reach more than 3.5 million in 2006 (*People's Daily*, 2003b; Zhou, 2005). For example, when Oracle sought to hire 23 new graduates for its software development centres in Beijing and Shenzhen in 2003, it was deluged with more than 4,800 applicants (Wharton, 2005a). Although applicants are always plentiful in China and foreign companies are particularly attractive employers, it is still a big challenge to find well-educated people for R&D activities. The quality of graduates lags behind the rapid increase in their quantity. In some industry areas, such as IT and telecommunications, the fight for top talent is becoming even fiercer due to the higher demand on the job market. As an interviewed CEO of a software company stated: 'We have a lot to do. If I could, I would hire 20 software developers immediately, but good people are very difficult to find.' The questions that arise here are what defines 'good people' and where does one find them? The next sections outline key recruitment criteria for R&D professionals in China and give an overview of the available channels for finding the right people.

HR selection

In order to find the right R&D staff when facing thousands of applicants, R&D and HR executives will have to develop jointly a list of key recruitment criteria and a technical job description. Table 6.3 lists the desired requirements for

R&D professionals, particularly with respect to local graduates, based on the interviews and literature.

The field research suggests that many general selection criteria for R&D professionals, such as language skills, soft skills, candidate potential or personality, seem to converge across industries, regardless of the MNC's origin. Yet the emphasis and must-have criteria will be influenced by the mission and personal preferences of the chief executives of the R&D units. Not surprisingly, research-oriented R&D activities, particularly in the pharmacy and chemical industries, have very high ratios of employed PhDs. In turn, masters and bachelors graduates form the majority of the staff in development-oriented R&D centres, and PhDs often have a project leader function.

In order to assess abilities and skills which graduation certificates and grades do not usually reveal, some MNCs often use typical assessment centre tools, including written tests (technical and non-technical), which are widespread in

Table 6.3 Key recruitment criteria for Chinese R&D professionals

Categories	Criteria
Education	Academic degree (top university) Self-education (extracurricular activities) Language skills (English)
R&D-related skills	Technical capability Practical experience (e.g. internships) Analytical and critical thinking Creativity
Soft skills	Communication skills Teamwork competence
Potential	Learning ability Flexibility
Personality	Proactive, independent, responsible Polite, honest, loyal

Source: Author; CH-ina (2005a)

China, to evaluate the candidate's suitability. Part of the interviews will be conducted in English. As the standard of education in China is relatively uniform, experienced R&D managers and HR executives mostly want to know if the candidate has made the best of his/her education and what kind of capability and skills he/she has gained outside the campus through extracurricular activities. Internships have become an important indicator for assessing and differentiating the ability of the graduates; however, these were only recently introduced as a mandatory part of study at Chinese universities and their quality varies greatly. In particular, internships within international companies are highly appreciated. It shows that the candidate has already gained familiarity with the international working environment and style. Moreover, such highly prized internship positions also reflect the quality of the candidate indirectly, as he/she had been selected from fierce competition. The findings of the last section indicate that MNCs have to pay particular attention to Chinese graduates' and young R&D professionals' potential deficits and particularities by defining key recruitment criteria. In order to assess the candidate's underlying potential, suitable evaluation modules such as a creativity test could be implemented.

Candidates with work experience, which is difficult to find in China, are mostly required for certain senior positions. MNCs find the 'global Chinese' a very attractive group of people. 'Global Chinese' is a generic term for three subgroups of Chinese people: mainland Chinese returnees with study and work experience abroad, Chinese from Greater China (i.e. Taiwan, Macao and Hong Kong) and overseas-born ethnic Chinese. Due to the huge immigrant Chinese community in the USA, US-based MNCs, more than, for example, European companies, can specifically benefit from scientists, engineers and managers with a Chinese heritage. There are three times more Chinese-born scientists with doctorates working in the USA than there are in the UK, the largest contributor of foreign-born scientists and engineers from the OECD countries (OECD, 2003a). The number of Chinese returnees (lyrically named *hai gui*[9]) working for foreign R&D labs has increased, keeping pace with the progress of the country's opening up and increasing opportunities. Many of them bridge the gaps experienced at senior employee levels. As explored by von Zedtwitz's study (2004), the initial expatriate R&D directors are increasingly being replaced by global Chinese, as they are in a better position to run R&D sites as integral elements of local scientific networks. Several of the interviewed directors of MNC R&D labs in China had quite similar career tracks. Chinese from the mainland who gained their PhD degree overseas and also had significant work experience abroad are especially attractive for positions at the senior level. The trend that sees increasing numbers of overseas Chinese scientists and engineers returning is not only due to the high appreciation of their work in China, but also because there are more opportunities for them and they can continue their careers with the advantage of being in a place that they call home.

It is not difficult to find global Chinese who have an overseas education in a specific field and can speak Chinese. However, they are also expected to serve as mediators and integrators. In reality, the overall requirements are therefore much higher than some Western HR executives realise. Several issues are often underestimated. First, it is obvious that the candidate needs to have the required technical expertise and knowledge in a specific field. But it should not be forgotten that he/she also needs to be able to communicate with Chinese researchers and staff about these issues in Chinese. Since most global Chinese have an overseas education and have spent many years abroad, some lack the ability to translate specific terms into Chinese. Second, they should be able to withstand the strong pressure and stress exerted by both their Western employer and the Chinese staff. Their Chinese colleagues will have noticeably higher expectations of them, which is often due to the large salary gap between them and the global Chinese. Third, the global Chinese need to act as human bridges and should be sensitive towards both Western and Chinese cultures. Ultimately, they are expected to handle intercultural conflicts. Although the global Chinese originate from the Chinese culture, it does not necessarily mean that they have the required sensitivities and skills to identify intercultural challenges, and to analyse and solve these kinds of problems. These issues should also be given attention when selecting personnel for these kinds of positions.

Recruiting channels

Today, HR executives in China can rely on a well-established portfolio of recruiting channels, such as public advertisements (newspapers, magazines), the internet, job fairs, campus recruiting, head hunters, governmental employment service providers[10] etc., to address potential candidates. MNCs generally use multiple channels simultaneously in their search for talent. The choice of recruiting channels largely depends on the size, industry, company policy, the MNC's previous experience and, of course, the type of candidate that the MNC requires.

Some best practices for recruiting R&D professionals in China have surfaced as a result of the field research, and are now discussed in more depth. Advertisements in newspapers or magazines, mostly in Chinese, are commonly used by large, well-known MNCs for their more general recruitment announcements, but are hardly ever targeted at R&D professionals. The importance of the internet in searching for staff is increasing rapidly in China. Not only are the postings of job offers on a corporate site relevant, but specialised online advertisement and recruiting services[11] now play a greater role and are particularly popular with young people who grew up with the internet. These are the channels MNCs will mostly use for R&D-related recruitment. If MNCs prefer inexperienced employees who are easier to train with respect to the corporate culture and business methods, campus recruiting is an effective way to access them. Through this form of recruitment, an MNC is also able to build a kind of 'brand' recognition by being present regularly. This is especially important for lesser-known MNCs. The prerequisite is that the company must have a clear understanding of which universities in China provide the best graduates in a particular scientific field. Some investigation must be carried out in this regard to ensure the recruitment drive's success. However, graduates from specific faculties that are well known nationally for their excellence have very good bargaining power. Not surprisingly, many foreign R&D centres have had positive experiences in hiring previous project members from university cooperation or internship programmes. With certain prior training, these recruits can start being productive much faster and the company has a chance to evaluate the candidates' potential and performance beforehand.

Due to China's strong cultural roots, one of the most important recruiting channels is the personal network. The well-known *guan xi* that governs business relations also influences people's employment process. A foreign company that has already gained experience with local staff often resorts to networking and referral, which is a very common Chinese HR practice. It is quite usual for companies to offer incentives to internal staff to recommend similarly skilled friends, former classmates and sometimes even relatives. This approach works particularly well within the Chinese context, as the quality of the referral has a strong impact on whether the referrer will gain or lose 'face' in the company.

In R&D, especially in high-tech areas, many interviewed companies hired their staff based on their *guan xi* with particular professors at specific universities.[12] This focused approach ensures the candidates have the qualifications that are specifically required by the company. One interviewee stated, 'The best R&D professionals are recommended by professors.' But these people are rare, as prominent professors usually have many industrial connections and much demand for their students. A solid and long-term relationship in R&D cooperation is often a precondition for potential recruitment. The more research-focused an R&D activity is, the more important this kind of talent-scouting approach.

Training and development

The discussion earlier in this chapter has shown that in order to leverage the high potential of young Chinese R&D professionals, training and development are necessary. As one interviewee pointed out: 'They are almost the "first-generation" talent in China. They do not have role models – most of their parents worked for state-owned companies.'

In general there are two sorts of training offered by MNC R&D centres in China. Technical training aims to close the gap between general and specific knowledge and practical ability which fresh graduates did not have the opportunity to develop during their studies. In certain R&D areas in IT and the semiconductor industry, newcomers can only be productive after at least several months of intensive training. In some extreme cases the training required will exceed one year. Experts from the headquarters or other international R&D labs will often be invited for intensive on-site training of local R&D professionals. Sometimes selected candidates with high potential will even be sent abroad to attend seminars and train. The second kind of training focuses on the development of the soft skills discussed earlier. Skill development training and workshops are usually offered to new employees at large MNCs. Unilever, for instance, offers new staff communication training (presentation skills etc.), project management, time management and language training.

MNC executives have different initiatives to improve their staff's skills and capability on a regular basis. For example, in order to improve employees' verbal English skills, one interviewed R&D centre encourages staff to speak English two days a week. Several of the interviewed centres have introduced weekly or monthly coaching opportunities, often lasting only one afternoon or even less. The topics covered at these events, which can take the form of mini-workshops, presentations or discussions, depend on what the supervisor thinks could be helpful for staff development, e.g. Western management theory and practice. It is also possible for employees to express their preference. This kind of regular training and development opportunities might be effective and useful to the staff, as the learned skills and knowledge can be directly implemented in their daily work. The efficiency of such training is generally higher than that of training that only occurs once or twice a year. Foreign managers should also ensure that they have a flexible mindset; in other words, they need to be sensitive to and integrate local culture into the coaching content and modify their ways of giving instructions. Before advising and making decisions, foreign supervisors should listen carefully to their Chinese staff (CH-ina, 2005a). The attitude that Western managers know better should be avoided.

Many R&D executives have quite convergent opinions regarding the question of how labs could increase local R&D professionals' poor individual initiative and lack of innovative mindsets. The most important factor in helping Chinese R&D staff to become more innovative and creative is a stimulating working environment (e.g. the physical setting of the workplace) and lab culture. The R&D lab needs to create a creativity-friendly atmosphere that influences the way of thinking. Most importantly, supervisors should adjust their leadership style accordingly. Employees should be encouraged to try new things, and should also know that they will not be blamed for showing individual initiative if mistakes are made.

An effective way to achieve accelerating learning curves by local R&D professionals is to send them abroad, often first to the corporate headquarters. ABB, for example, regularly sends its R&D staff to the company's global R&D headquarters in Switzerland. According to Unilever, more than half of the company's R&D staff have had training, education and work experience abroad. An international posting is an exciting experience, especially for young professionals, as most of them have never been outside China. It is therefore also a very effective means to enhance an employee's motivation and loyalty. At the same time, the increasing internationalisation of R&D and the Chinese labs' soaring importance within global R&D networks offer more opportunities than ever before for Chinese R&D staff to work at an international level. Several MNCs offer job rotation programmes or project participation at an international scale, in order to increase local R&D professionals' ability to work in an international environment and build up international networks, which are essential for the increasing number of cross-border research projects. Sometimes the stay abroad extends for as long as one or two years. Not only are the language skills of these 'Chinese expatriates' improved by their international work experience, but also their understanding of the parent company's culture and their international colleagues' working style is improved. Local R&D staff who are sent abroad for training and projects have often been identified as employees with high potential and are considered as future candidates for key positions. They help MNCs to intensify further the localisation of international R&D labs' management. Despite the high cost of overseas training and staff development, it has become a regular practice in some research labs, as the expected return is much higher. Some foreign R&D labs often use the strategy of 'train the trainer first'. The first batch of hired R&D staff will mostly be sent overseas for training. They then take over the coaching of the second-generation staff. An interviewed MNC encountered an obstacle in this respect, in that within two years the majority of the first batch of overseas-trained 'trainers' had left. Such a generation-focused staff training approach may therefore not have its desired effect due to the relatively high employee turnover rate in China (discussed later).

As mentioned in previous sections, Chinese returnees who have international experience could be suitable candidates to fill key positions. Nevertheless, the field research also revealed that some Chinese returnees do not integrate well despite expectations in this regard, thus some companies would rather hire local people and train them for a few years to catch up. Returnees will thus be increasingly treated like other local staff members. Those without directly applicable experience will receive the same initial training as the local R&D professionals. This implies that without any customisation to and preparation for the 'old-new environment', the overseas or global Chinese may have difficulty in doing their appointed tasks and fulfilling their role as 'leveragers' and 'mediators' in China, and their impact will be called into question. Proper training and preparation are thus necessary for global Chinese with high potential to enable them to act as 'human bridges' or 'glocal Chinese'.[13]

The field research revealed that some foreign R&D labs are currently facing a lack of qualified local managers and project leaders – those who have both a technical background and managerial competence. In order to cope with the demand for R&D managers, Unilever offers a dual-track system to enable suitable candidates with a technical background to strive for the manager track. A variety of training in management and soft skills is provided. However, the manager track is limited in respect of time. After every two years the candidates' suitability is reassessed. Regular performance evaluations and talks with a supervisor ensure that the manager track is right for the candidates and that they are able to meet the requirements.

Overall, Chinese R&D professionals still reveal several deficits when compared with their international peers, but their underlying potential is huge. For international R&D labs to leverage this potential, training and staff development are therefore of particular importance. In general, Chinese R&D professionals are keen to receive training, as they are very aware of their educational system's limitations and are thus highly motivated to learn marketable skills. However, training and staff development can only be successful if the needs and gaps in practice are adequately identified and properly addressed in accordance with the long-term corporate strategy.[14]

Evaluation and remuneration

Evaluation

An effective evaluation approach is instrumental to appropriate staffing, objective remuneration and proper staff development and training. The field research revealed that MNC practices in evaluating their Chinese R&D staff follow one of two different evaluation streams. Some MNCs still rely on the traditional top-down approach of employee evaluation, i.e. the supervisor is mostly the only one who evaluates the subordinate, using a checklist of different aspects prepared by the HR department. One interviewed foreign R&D lab, for example, places strong emphasis on the individual employee's technical achievement and performance within a project as well as the projects he/she was assigned to do as a whole (e.g. meeting deadlines, the quality and cost target of the results). Several other factors, such as efficiency, the project's difficulty and the required technical skills (e.g. new software and hardware platforms involved) as well as the project's innovativeness and staff, will be taken into account. Furthermore, a range of soft factors such as communication skills, teamwork, leadership skills (for employees with line functions) etc. will be evaluated. Most local Chinese R&D professionals are familiar with this kind of top-down assessment approach, as they have been evaluated in this way almost throughout their life. However, the second stream of practice approaches the modern understanding of staff evaluation, i.e. a 360° evaluation method. With this approach, besides the purely technical performance evaluation, employees are not solely evaluated by the supervisor but also by colleagues, subordinates and internal and external partners. Above all, he/she must also evaluate him- or herself. Most of the interviewed Chinese R&D staff had never experienced this kind of evaluation before and reveal a positive attitude towards it. However, for the previously discussed cultural reasons, the first difficulty one could encounter is that local Chinese employees are reluctant to provide critical and honest comments regarding their team colleagues (in-group/out-group thinking). They tend to provide more positive appraisal. Secondly, in terms of self-evaluation, many are inclined to be over-critical – a result of the emphasis placed on being humble, a central virtue in Chinese culture. Keeping these issues in mind, Western HR and R&D executives need to adjust their usual assessment approach in respect of the underlying cultural bias.

Remuneration

The remuneration practice of foreign R&D labs in China is often in accordance with the parent company's general HRM policy. Compensation policies, as explained by the majority of interviewed companies, are based on what is called the magical triangle (Hilb, 2004): personal determinants (educational background, experience and individual performance), HR supply determinants (availability of personnel with similar skills and experience on the market) and corporate determinants (company performance or, in the case of R&D labs, determinants such as completed projects, number of innovations, patent outputs, publications etc.).[15]

As the findings of research phase I indicate, one of the motivations for MNCs installing R&D activities in China is the cost, particularly the low cost of local labour. However, during the last few years personnel costs in China have steadily increased, particularly regarding specialised professionals[16] in first-tier cities such as Beijing and Shanghai. A foreign company's local line manager in Beijing or Shanghai earns US$27,000–32,000 (total annual salary plus bonus); senior managers receive US$46,000–54,000; and top executives can reach US$80,000–90,000 or more (*The Economist*, 2005c). On average these salary levels increase annually by 6–10 per cent, with inflation of 2 per cent (Wharton, 2005b). The picture is totally different if one looks at the wage level of fresh university graduates in China. An online study undertaken by the *Beijing Morning Post* revealed that 41.5 per cent of them expected starting monthly salaries of RMB2,000–2,999 (US$247–371), 34.6 per cent expected RMB1,000–1,999 (US$124–247) and only 21.5 per cent said they expected more than RMB3,000 (US$371) a month (Zhou, 2005). An interviewed R&D executive maintained that although R&D staff's annual salary growth in, for example, Shanghai was around 10 per cent, the wage level is still low compared with that in the triad regions. Some executives point out that local staff's performance and efficiency are also lower, especially in terms of innovation output, than those of their European or American peers, but that by calculating the performance-cost ratio, Chinese R&D professionals have a clear 'advantage'.

As pointed out in the previous section, Chinese returnees could play an important role in MNCs' success in China. Simultaneously, those returnees' remuneration has always been a sensitive topic. The field research revealed one interesting development: in general, Chinese returnees' salary level is much higher than the local level, but lower than in the West or that of non-Chinese expatriates. The salary gap is particularly big between locals and returnees with higher qualifications, e.g. in one of the interviewed R&D labs, a local with a PhD earns around RMB100,000 per year while a returnee with a PhD earns RMB200,000–250,000 a year. Only a few top executives of Chinese origin who have wide overseas as well as local experience will receive expatriate treatment.

Recently, due to many MNCs' widespread staff localisation policy[17] and changes in HR supply determinants, i.e. a soaring number of returnees and the local graduates' improved qualifications, the market value of less-specialised overseas-trained Chinese has started to decrease. Some MNCs have even introduced equal treatment and remuneration for local employees and Chinese returnees. In other words, the returnee starts with a salary similar to that of his local colleagues (local contract). Besides the changing HR market conditions, one important rationale behind this kind of policy is that although the returnees have been educated overseas (usually an indicator of a better qualification), it is their actual performance that counts. Moreover, unequal compensation based on a differentiation between local and overseas educational backgrounds could affect local employees' motivation. The side-effect of this policy is that MNCs will have some difficulty in attracting returnees, especially the talented ones. Many returnees have high expectations in terms of compensation and status from a mainland job. From their perspective, the new policy is unfair because compared to local graduates they have invested much more in their education in terms of effort, money and time. There is also a 'face' issue behind returnees' attitude, since their social environment's expectation is very high. However, executives' experiences have shown that some Chinese returnees who have started at the same salary level as the locals are able to increase their remuneration rapidly because their performance is considerably better than that of the local employees. The years of experience abroad clearly result in high-quality work. The implication to be drawn here is that a transparent and performance-oriented compensation system and, indeed, clear communication of career development opportunities are key to leveraging returnees' motivation. From the HR selection perspective, people who accept this approach also tend to excel.

The many irregularities of the HR market in China, e.g. its poaching practice, are responsible for the job-hopping that has become the usual way for professionals to increase their income (see below). As verified by an interviewee, the same job will sometimes differ greatly in remuneration due to job-hopping. In certain areas specialists are already as expensive as their international peers. Although this kind of situation follows the free market economy's rule that demand and supply dictate the price, it is a tremendous challenge for companies to retain their people.

Retention

The phenomenon of high staff turnover rate

This section tackles the source of many HR executive headaches in China: retaining talent. Almost all interviewees underlined that it is a pressing concern for MNCs to acquire good people, and even more difficult to retain them due to the widespread job-hopping mentality of the young professionals. For example, a research study revealed that one in ten executives in Shenzhen, one of the first special economic zones (SEZs) in China, changed jobs in 2004, and one in 12 in Beijing. The average employee turnover rate in the country was estimated as being 11.3 per cent[18] in 2004, up from 8.3 per cent in 2001 (Wharton, 2005b). The area of R&D is not immune to the trend.

Before employers try to retain their employees, they need to know the reasons for staff leaving. In general, there could be a series of reasons, e.g. a limited career perspective, the corporate culture, payment, discontent with the management style or inadequate training and development opportunities. Within the Chinese context, payment is widely viewed as having the most significant impact on the HR turnover rate. A supply-demand disparity regarding HR in China and socio-psychological reasons jointly explain this phenomenon. The rapid economic development has accelerated the demand for young talent and specialised professionals with a good command of English who are ideal candidates for MNCs and large domestic companies. Due to foreign companies' rapid inflow into China after its WTO accession and the fast-growing private sector, the competition for employees with high potential is becoming even fiercer. Foreign companies that have just entered China are aggressively trying to hire people away who have been trained by other MNCs in order to jump-start their activities. This has led some MNCs to start to seek new HR markets outside Beijing and Shanghai, e.g. in second- and third-tier cities or in the western regions. Infineon Technology, for example, in September 2005 established a new R&D centre in Xian, a hub in China's western region, with more than 350 researchers and engineers. Future expansion has already been planned. Despite the fact that China has continuously increased its number of university places, the education system is not able to supply enough capable graduates to meet the rising demand.

There is also a socio-psychological issue behind young, talented people's low degree of loyalty. Before the economic opening, China was a mono-option society. A young person's career path was predefined and mostly controlled by the state. The young people of today, at least in urban areas, have grown up in a multi-option society, in which everyone is responsible for his/her career advancement and is offered multiple choices for self-actualisation. People have seen that it is quite possible to become wealthy within a short period of time, and young people have grown up with the famous saying of Deng Xiaoping, the initiator of the reform to open China, 'to become rich is glorious'. The longing for wealth is stronger in the society than ever before. It is thus obvious that the psychological barriers against hopping from one job to the next are extremely low.

A good number of talented employees have left their employers for an overseas education in order to achieve a higher 'market value'. During their university studies, many noticed that the Chinese education system could not equip them with the necessary knowledge and competency they would need for their desired career. Working for an MNC for a while before going abroad is considered as a kind of preparation. This not only provides financial support, but the MNCs are able to give excellent training to allow them to become more familiar with the international context.

Leveraging the retention challenge

It is known that retention is much cheaper than recruitment; but to realise retention effectively remains a huge challenge for HR and R&D executives in China. Compensation is often considered one of the most effective ways to retain staff within the current situation. According to the interviewed executives, R&D professionals are not as payment-sensitive as others, e.g. those working in marketing and sales. Nevertheless, it has become more complex and challenging to offer a competitive compensation package. Besides salary and performance-related bonuses, a series of perks such as housing allowances, medical and accident insurance and supplementary pension plans now form part of an attractive compensation package for persons with high potential who work for foreign companies in China (see also Wharton, 2005b).[19] Due to the dynamism of the HR market, it has become increasingly crucial for HR executives to be well informed about current developments in staff remuneration.

Surely an *attractive payment and incentive package* could to a certain degree ensure that an employee remains loyal to an employer? No, without paying attention to other aspects, a larger wage increase will not ensure an employee's attachment to a company. This is clearly confirmed by a survey carried out among 144 employees[20] of Swiss companies in the machinery and electronics industries in the Shanghai area in 2004 and 2005 (CH-ina, 2005a). Although ranked just after 'limited career perspective', 'compensation and benefits' has the strongest impact on an employee's decision to leave; but it is, interestingly, only ranked ninth among all the influential factors for employees to stay. In other words, employees' decisions to stay with their employer are not mainly driven by monetary incentives.

The interviews largely confirm the finding of this survey. Showing people with high potential in R&D a *career perspective* is particularly relevant in retaining them. *Training and career development programmes and plans* include participation in corporate (international) training programmes and/or attendance at university training courses. Several large MNCs, such as General Electric and Hewlett-Packard, maintain an in-house 'university'. Large MNCs' experiences have shown that employees who receive training tend to stay longer (Wharton, 2005b: 5), while it is key to match individual as well as corporate needs. In line with findings by Melvin (2000), several companies complained that they lost their brightest people to postgraduate study abroad. Instead of just letting them go, a promising approach could be to offer talented employees tuition reimbursement in exchange for remaining with the company for a certain number of years after graduation.

Keeping top talent also means that *their skills need to be regularly challenged*. A higher degree of employee participation in decision-making processes could be helpful, and also continuous *access to new knowledge* needs could be guaranteed. The field research revealed that the pattern of job change is likely to be linked to the learning curve. The foreign R&D labs interviewed have brought to light that newly hired employees are most likely to change their jobs between the second and third years. After this period of time their learning curves usually start to slow down. For R&D professionals in research-focused areas, however, the learning process could take longer. Using the pharmaceutical industry as an example, interviewees maintained that the 'risk period' of losing people starts much later in this industry than in other industries, as the R&D process for a medical compound lasts longer than in other industries. High-potential Chinese R&D staff are becoming more sophisticated, and are no longer satisfied with less-challenging tasks. However, the field research also recognised a certain reluctance by MNCs to extend their Chinese R&D labs' mission. On the one hand, the headquarters might start doubting the local team's capability and have IP-related concerns. On the other, internal resistance might be rooted in fellow international R&D units' worries about losing their internal charter to China. If this is the case, the Chinese lab's executives have to convince the HQ to develop more trust in the Chinese R&D professionals and expand their mission in order to meet their fast-growing aspirations. An independent expert even sees it as an obligation for MNCs to let local R&D staff grow professionally.

A *friendly and enjoyable working environment* needs to be mentioned as a further factor strengthening employees' attachment to a company. This embraces not only physical aspects of the workplace, but also soft aspects such as giving competent people the freedom to manage their own work and flexible company regulations, e.g. regarding working hours and meal times. The answer is to make a difference and create a *familiar atmosphere* for the staff to make them feel comfortable. Besides this, several interviewees believe that *the company's promising development* helps to increase employees' confidence in their own career. This issue is especially relevant for small MNCs.

An issue frequently underestimated by Western executives is the role of *personal relationships* in retaining employees. In general, the personal bond between a superior and his/her subordinates is much stronger than in the West. It is common in China that when a manager leaves a company, he/she is able to take many other employees with him/her. Western managers often complain that they do not have the chance to recognise their people's mood: 'Once they come to talk to you, they have already made their decision to go.' This is because Chinese subordinates are not used to talking to their superiors unless they feel they have established a close personal relationship with them. Contrary to the Western working culture, teams and their team leaders in

China not only spend a lot of time together at work, but also after work. Leisure events and dinners will be organised regularly to strengthen the in-group feeling and get to know one another better. This is also an opportunity for the team leader to talk to his/her subordinates informally and increase his/her relationships with the team. An interviewed Chinese manager revealed that he meets his team and their families at least once a month for leisure activities. In an in-group-oriented culture that simultaneously has a high power distance,[21] the superior often uses these kinds of means to maintain a close relationship with subordinates and become well informed about their thinking. In a more individualistic Western context such events occur perhaps only once a year, since people do not miss this type of communication channel and almost everything can be discussed at work. Thus, Western expatriates often have the attitude that 'Work is work, and I do not want to spend my precious spare time with my colleagues.' They prefer to spend their leisure time with their families and the expatriate network, rather than with their local staff. It is also not helpful that many expatriates are assigned to China for only a limited period (usually three to five years), as they will be transferred at just about the time that they have gained adequate experience of the Chinese environment and have built up close personal relationships with the local staff. For this reason, some MNCs try to convince expatriates to stay longer, while others increase the number of managers of Chinese origin.

Although a healthy HR fluctuation stimulates the inflow of new ideas and knowledge, fundamental to innovation, a high staff turnover rate could endanger the sustainability and success of an R&D activity. Simultaneously, loyalty is an effective means to prevent IP leakage, which has increasingly gained importance against the background of challenging IP protection in China (see Chapter 7).

Nevertheless, long-lasting loyalty and retention of employees cannot be 'bought' through higher salaries: a well-balanced mixture of different factors is required.

Conclusion

In this chapter the discussion on managing R&D staff in China has first revealed the strengths, deficits and particularities of young professionals and graduates in China. Besides solid technical know-how and theoretical education, Chinese R&D professionals are particularly motivated, diligent and have high ability and a willingness to acquire new knowledge. Simultaneously, due to the weaknesses of the Chinese education system and cultural biases, many reveal particular deficits. In R&D-related areas there is a lack of innovativeness, independent working ability, hands-on capability and complementary skills such as oral English, presentation and teamwork ability. Nevertheless, a context-adjusted management practice and proper training and coaching opportunities will allow MNCs to leverage Chinese R&D staff's huge underlying capability. Recruiting practices for R&D professionals in China not only entail hard criteria such as education and experience, but considering the professionals' hidden potential as one of the decisive issues. The discussion has further shown that, analogous to China's business world, recruiting Chinese talent could also be *guan xi* based. Due to the risk of high staff turnover, staff retention has become increasingly significant. The central implication to be drawn from the research is that although salary is important, Chinese with high potential will not remain purely for a high salary. Career development opportunities, challenging tasks, atmosphere at work and a close personal relationship with superiors are keys to retaining them in the long run.

From a strategic perspective, the management staff of a foreign R&D lab will probably have the strongest impact on its success. From a competence-based view, this phenomenon can be described by means of the R&D unit's absorptive and multiplicative capacity (see Levinthal and Cohen, 1990). Both capacities determine a firm's capability to link its peripheral units to the external environment. A higher absorptive capacity allows an MNC to have a better knowledge-sourcing capability in the local community, while a higher multiplicative capacity improves the firm's capability to link the peripheral unit's output efficiently to the MNC's global operations (R&D effectiveness and efficiency, and technology transfer). In a global R&D unit, expatriates and global Chinese fulfil the two functions of absorbing local skills and knowledge, and multiplying the output to the MNC's operations. Expatriates are usually in charge of the multiplicative capacity as they are more familiar with the parent company's organisational network, culture and strategic focus. Global Chinese are often assigned to be responsible for the absorptive function, since they should be more familiar with the local environment. However, the research has established that more and more global Chinese are taking over expatriates' functions, especially those Chinese who have worked overseas for MNCs for a long time.

In the long run, both multiplicative and absorptive capacities could be leveraged by the global Chinese. The cost-saving factor is only a minor reason for MNCs to adopt this policy, as some high-profile global Chinese are paid almost as highly as expatriates. It is rather because global Chinese are in a better position to take on the expatriates' function, i.e. the multiplicative capacity, than vice versa. For most expatriates it is quite difficult to become integrated into important local communities outside the company, but this is a necessity for an absorptive capability to match that of their global Chinese colleagues. Culture and language remain the major challenges.[22] Keeping this in mind, MNCs need to pay greater attention to identifying and attracting appropriate candidates who can handle these challenging tasks. They need to be experts in the technical field, familiar with the organisation and share the vision of the parent company, have enough leadership quality to manage and train young local talent, show sufficient entrepreneurial spirit to lead the R&D centre and also have the ability to span the cultural boundaries between East and West. A personality that could meet all these criteria resembles the role-model manager 'glocalpreneur' (Hilb, 2002). Yet it will prove difficult for MNCs to find candidates who meet all the criteria.[23] The most effective strategy might be to develop promising individuals within the company.

Notes

1 '[M]ost local Chinese managers of foreign-owned operations are aged 25–40, with the vast majority aged 33–36. Applicants over the age of 40 – who were in their teens and twenties during the Cultural Revolution – tend to be overlooked for management positions' (McComb, 1999: 31).

2 In this context, the term 'young people' relates solely to China's urban population.

3 The average educational level of China's graduates is still low, although Chinese high-school graduates reveal excellent knowledge and scores compared with their international peers. The two weak spots of the higher education system are the faculty and lack of financial means.

4 Some of the deficits recognised by foreign managers are

'deficits' as measured according to Western culture and business practice. The author has consciously added 'particularities' in order to provide more objectivity and perspective, since strong cultural influences are involved.

5 'Confucianism expected everyone to strictly observe his or her prescribed position in the different social strata in order to maintain the "perfect social order" into perpetuity' (Liu, 2003: 176).

6 For example, certain development-oriented activities in biotech or chemical industry laboratories are based on diligence rather than on creativity and own initiative.

7 Respect for age and hierarchical position is also a reason why Chinese people are less critical towards their superiors.

8 For example, since Motorola established the first foreign-invested R&D centre in 1999, it has hired a total of 1,300 engineers. Philips has more than 700 R&D staff and will increase this to 1,300 over the next two years. GE's new global research centre in Shanghai was formally opened in 2003 and hired 500 researchers; it was expected to employ 1,200 by the end of 2005 (UNCTAD, 2005).

9 *Hai gui* means 'those returned from the sea', but returnees are currently widely called 'sea turtles' in China, due to the identical phonetics of both expressions.

10 The most prominent is the Beijing Foreign Enterprises Services Corporation, known as FESCO. It almost has a monopoly in helping foreign representative offices handle HRM issues.

11 For example, www.51job.com (a phonetic word play on 'I want job' in Chinese) or www.zhaopin.com ('recruiting').

12 One high-tech company's newly established R&D centre has hired almost half of the entire graduate class in a specific branch of study at a leading university in Shanghai, thanks to its good connections with the professors.

13 Inspired by the model of 'glocalpreneur' (Hilb, 2002). 'Glocalisation' originated from Japanese business practices in the 1980s. It was first popularised by Robertson (1995).

14 See, for instance, Hilb (2004: 91–8) for an integrated approach to staff development.

15 Usually individual and group performance directly determine the size of an employee's bonus and the remaining factors

influence an R&D professional's basic salary. In one interviewed company, the bonus can be up to 30 per cent of the remuneration.

16 Such as, for example, experienced marketing managers and finance or accountancy specialists with a good command of English.

17 There are two major reasons for MNCs to promote staff localisation. First, local managers and staff have the natural advantage of knowing the language and the culture. Second, expatriates are always very expensive. 'A typical expat manager may earn $200,000 to $300,000 a year in base salary and receive a $10,000 housing allowance each month. His firm may also give him $20,000 to $40,000 annually to educate his children in private schools, pay $20,000 to $40,000 a year to cover the cost of trips back home, and also pay his taxes' (Wharton, 2005b: 5).

18 Some smaller firms face a turnover as high as 30 per cent (Wharton, 2005b).

19 It is also traditional in China for an employer to provide food, transportation and even housing for employees.

20 Including managers, supervisors and workers (CH-ina, 2005a).

21 Power distance is defined as 'the extent to which the less powerful members of institutions and organisations within a country, expect and accept that power is distributed unequally' (Hofstede, 1991: 28). Chinese societies are characterised by a significantly high power distance.

22 It is worth underlining that the increasing cultural awareness among young Chinese professionals and MNC expatriates who work together will surely lead to a gradual convergence of their cultures.

23 However, it is not impossible to find this kind of people. As far as the author can judge, several 'glocalpreneur' type of Chinese executives were interviewed during the field research.

7

Managing intellectual property rights

Managing intellectual property rights (IPRs) is one of the major concerns for many MNCs doing business in China. The reason is obvious: the Chinese environment has comparatively serious deficits in different areas of IPR protection. Despite China's WTO accession in 2001, IPR infringement has remained unacceptably high. For instance, more than 90 per cent of software used in China is counterfeit (Weinstein and Fernandez, 2004). According to 'Special 301 report of the United States Trade Representative', the value of Chinese trademark counterfeit products entering the US market increased 47 per cent from US$94 million to US$134 million in 2004, accounting for 67 per cent of all US Customs IPR seizures in that year. Moreover, foreign pharmaceutical companies in China lost annual revenues of 10–15 per cent due to counterfeit products (US Trade Representative Office, 2005). Without exception, technology-intensive industries are implicated. In recent years several IPR disputes in the high-tech sector have received increased attention in China (Table 7.1). It is thus no surprise that the results of the first research phase clearly indicated that IPR issues are considered one of R&D management's biggest challenges in China. For example, ABB maintains that illegitimate technology transfer through

Table 7.1 Recent prominent IPR disputes within the high-tech sector of China

Year	Case	Source
2005	Intel accused Shenzhen-based high-tech company of copyright infringement and asked for US$8 million in compensation	*People's Daily* (2005c, 2005d)
2004	Two American enterprises claimed Chinese DVD chips infringed their patent and requested US International Trade Commission to initiate an investigation	*China Daily* (2005)
2003	TSMC sued Shanghai-based Semiconductor Manufacturing International Corporation (SMIC) for stealing trade secrets and asked for US$1.7 million damages award	*People's Daily* (2003c)
2003	Cisco Systems sued Huawei for infringing its process and method patents, trademark and trade secrets	*China Daily* (2004b)
2002	Huawei Technology sued three ex-employees for IP theft	*People's Daily* (2004b)

licensed parties, which leads to new sources of revenue being withheld, is one of the major IPR infringements the company has to face in the country.

At present technology is of strategic importance for many companies to compete successfully in the market. IP is therefore one of the most valuable corporate assets, especially for technology-intensive industry sectors. Yet MNCs in China are facing a dilemma. On the one hand, given the concerns regarding IP infringement, MNCs are more reluctant to bring cutting-edge technologies to the Chinese market. According to the Swiss China Survey,[1] even successful firms can only succeed in every second patent-right-infringement case on average (CH-ina, 2005b). On the other hand, fiercer competition demands continuous technological upgrades in order to remain competitive. Managing IP appropriately within the Chinese environment has become the centrepiece of solid corporate development for high-tech companies.

This chapter has three main focuses. First, it reviews the literature on IP in China to give a picture of the existing stock of knowledge. Based on previous research results and identified literature gaps, research questions will be formulated. Second, the background and existing IPR protection framework in the country are revealed. Third, based on the results of field research, identifiable major challenges in managing IPR in China are discussed. The chapter provides an integrative approach for dealing with IPR in China as a whole. Given the focus of the research on R&D management, IPR issues under investigation will be primarily tackled from technology-intensive companies' perspective.

The environment of IPR protection in China

The IPR protection environment in China is evolving rapidly. This section first explains the short history of IP legislation in China. Second, the invisible environment, i.e. the cultural and social influences in China on IPR protection, is discussed. Lastly, official endeavours to strengthen IPR protection in China will be depicted.

The development of IP legislation in China

Compared to more than 200 years of IP legislative history in the West, the development of IP legislation in China is in its infancy. Before China introduced reforms and opened up to the world, there were no specific laws or systems regulating

IP. However, within 20 years China established a comparatively mature IP regime (Table 7.2). The Trademark Law, the first of its kind, was promulgated in 1982. This is considered the first milestone regarding China's IP legislative reform. To meet the needs of economic development and scientific advancement, comprehensive legislation was undertaken in the following two decades. A series of new IP laws have thus been promulgated in China for the first time.

China's WTO accession in 2001 initiated a new round of comprehensive revision of the legislation, as all WTO members have to adjust their laws and enforcement procedures in compliance with the TRIPS agreement (Table 7.3). Consequently, a number of articles in the Trademark, Patent and Copyright Laws were deleted, rewritten and created around 2001. In addition, the rapid development of high-tech industries, such as IT and semiconductors, required China to adopt separate regulations to tackle new types of IP. Hence, several special regulations such as the Regulation on the Protection of IC Layout Designs and

Table 7.2 IP legislation in China since 1982

Year	IP legislation steps
1982	Trademark Law first enacted
1985	Patent Law first enacted
1991	Copyright Law and Implementing Regulations enacted
1993	Patent Law amended Implementing Regulations under Patent Law enacted
2001	Patent, Trademark and Copyright Laws all amended in response to WTO accession on 11 December 2001 Regulations for Protection of the Design of Integrated Circuits enacted
2002	Regulations for Protection of Computer Software enacted
2003	Implementing Regulations under Trademark Law enacted

Source: Based on Weinstein and Fernandez (2004)

Table 7.3 Important IPR legislative improvements after China's WTO accession

Area of improvement	Current status	TRIPS compliance
Patent term	Changed from 15 to 20 years through 1993 amendment	Compliant
Scope of protection	1992 amendment allowed protection of chemical and pharmaceutical products	Compliant
Exclusive rights	1992 and 2001 amendments added 'offering for sale' and 'right of import' to list	Compliant
Burden of proof in case of patent infringement	Burden to probe non-infringement is shifted to defendant	Compliant
Judicial review	Now available to review decisions of administrative board on re-examination, cancellation and revocation of patents	Compliant
Innocent infringer	Amendment clarified such that 'innocent infringer' defence is only relevant when calculating damages	Compliant
Compulsory licensing	Clarification of situations in which compulsory licence can be obtained	Compensation can be further clarified
Preliminary injunction	Now available	Depending on implementation
Statutory damages	Up to RMB500,000 (approx. US$60,000) when actual damages cannot be assessed	May not be enough of a deterrent

Source: Based on Weinstein and Fernandez (2004)

Regulation on Software Protection have been promulgated to meet the TRIPS standard (Weinstein and Fernandez, 2004).

In just two decades China has made remarkable progress in IP legislation, if one takes into consideration that the country started from zero.[2] After two rounds of revision,

Chinese IP legislation is comparatively mature and has almost reached the standard required by the TRIPS agreement. However, many researchers believe there is still a long way to go before China's IP laws are fully compliant with TRIPS.

The social awareness of IP

The general awareness of IPRs in China is noticeably weaker than in developed countries. The lack of an IP legal foundation in the past is one important reason; but the 'people factor' is often neglected. The people in China have experienced the transition from a planned to an open-market economy within a relatively short period of time. Many of the country's regulations have been changed fundamentally. Some of them can be easily adopted, others are more difficult to put into practice. This is especially true of areas in which the new regulations require a new way of thinking that is no longer in accordance with the cultural and social roots that invisibly influence how people think and act towards IPRs.

For over 2,500 years Confucianism has played a decisive role in shaping Chinese people's and society's mentality. Traditional Chinese thinking is not familiar with ownership of an immaterial good as an individual right; in fact, it strongly emphasises learning by copying, with a teacher being regarded as praiseworthy if a student can perfectly reproduce his work. Consequently, copying is widely accepted in Chinese society (Alford, 1995; Wingrove, 1995). The Chinese saying, 'to steal a book is an elegant offence', clearly reflects the traditional attitude towards IP. In contrast to the West, 'the copying of works of almost any kind has for centuries been regarded as honourable and necessary' (Wingrove, 1995: 6). Hence, IP infringement is regarded as a very minor incident from the traditional perspective.

More than four decades of communism impacted on the Chinese understanding of IP even more. Before China adopted reforms and opened its doors, property of any kind, such as intellectual work, was viewed as public assets that could be used or shared for the good of the community almost without the specific property holder's authorisation. This period increased the gap between social awareness and the concept of IP. However, with the rising importance of private property since China's economic reform, the Chinese people, especially business people, have started to acknowledge the general idea behind IPRs. Although IP law is generally considered far from important in most people's daily life, the younger, educated generation's IPR consciousness is increasing.

The governmental perspective on IPR protection

Over the years the Chinese government has clearly come to recognise the importance of IPR protection for the country's economy and reaching its ambitious goals.[3] A solid IPR protection system is not only a key element to attract technology-intensive FDI and bring in new cutting-edge technologies, it is also essential for domestic industries' sustainable growth and the further development of the national innovation system, which is the foundation of a knowledge-based economy. In recent years a series of nationwide campaigns have been launched to stop IP infringement and indicate this topic's importance for the authorities. For instance, the State Council has set up the

IPR Protection Working Group with China's vice premier as its director. In order to intensify efforts to combat IPR infringement, the State Council launched a year-long specialised campaign throughout the country in November 2004. To date the campaign has led to more than 1,000 IPR violation cases with a total value of over RMB860 million and involving more than 2,600 arrests (State Office of Intellectual Property Protection, 2005a). Temporary limited actions alone cannot, however, generate sustainable improvement in IPR protection in China, thus the government has also introduced educational campaigns through mass-media channels such as television and the internet. This kind of mass education could play an important role in establishing new customs and practices regarding IP in the society, if it is consistent and ongoing.

The major IPR protection efforts have been primarily concentrated on two kinds of IP infringements: trademarks and copyright. However, for the high-tech industries, IP infringements regarding patents and trade secrets are more relevant. In these areas IPR protection can be effectively achieved through strong enforcement measures case by case. Nevertheless, the officially initiated mass campaigns not only serve as a means of mass education, but also convey an important message to the world: China does care about IPR protection and is making steady progress.

IPR protection systems in China

While the last section depicted the IP protection environment in China, the following sections tackle several focal issues for foreign R&D activities: how the current IPR protection system functions and how IPR protection can be exercised with the help of the available means. First comes an explanation of what protectable IPRs there are in China and their applicability. Thereafter, key IP administration organs and their responsibilities are discussed in brief. Lastly, different types of IP dispute settlements are explained and the implications for IP strategy discussed.

Protectable IPRs in China

Broadly, four types of IPR enjoy legal protection in China: patents, copyrights, trade secrets and trademarks. Due to specific industries' need for IP protection, China has also introduced exclusive protection rights for computer software and IC layout designs.

Patents

In order to make the right decision about which patent type should be chosen during the filing process, a company needs to balance the economic value of the patent against infringement risks and economic damage to the company. China's Patent Law[4] differentiates between three types of patents: *invention*[5] patents; *utility model*[6] patents; and *design*[7] patents. As limited by the Patent Law, only technical solutions with industrial applicability and scope regarding products and processes can be patented. Considering the short life cycle that utility models and designs may have, they are not subject to substantive examination and are protected for shorter periods than invention patents. Generally, the applicant can obtain the patent in China within a year. In contrast, the examination of an invention requires a more thorough procedure, and it usually takes four to five years to issue an invention patent (Chen, 2005). Of all the protectable IPRs in China, patents have the strictest examination criteria and provide the strongest protection due to their long protection term. The patent application process in China is depicted in Figure 7.1. In order for the patent's protection to have a solid foundation, a thorough search and analysis of existing competitors' patents in China is a crucial step to be completed before starting the filing process. Moreover, patents to be filed

Figure 7.1 **Patent application process in China**

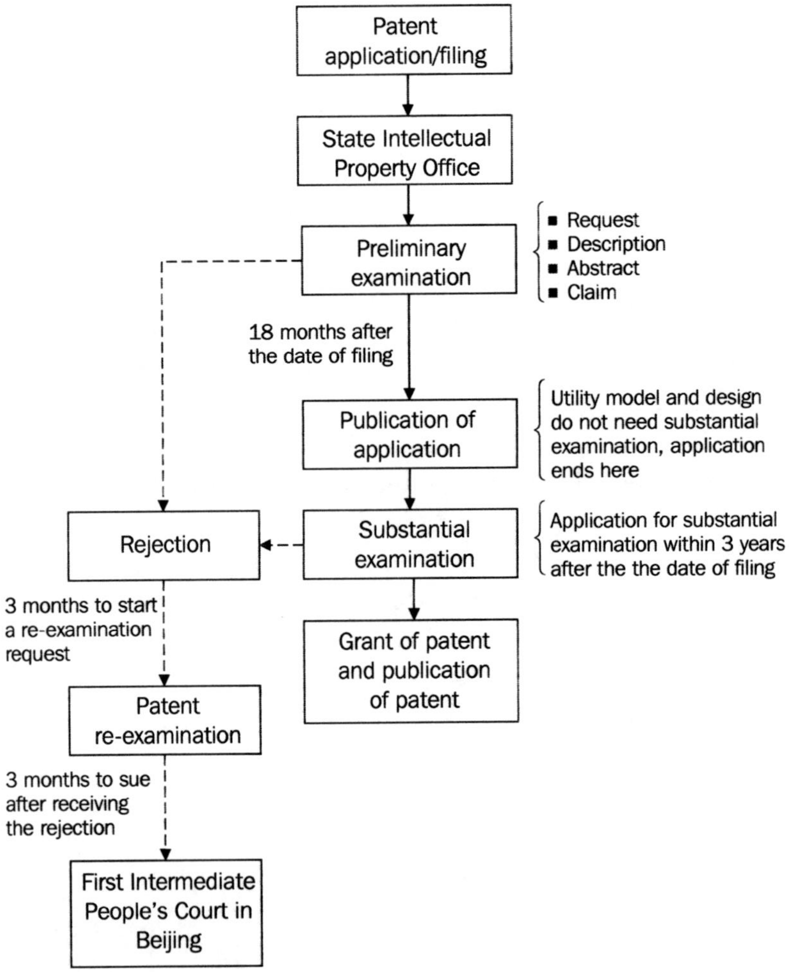

Source: Based on the PRC Patent Law

should be in line with the company's own corporate IP portfolio strategy. Applicants are required to submit a request,[8] description,[9] abstract[10] and claim form to the State Intellectual Property Office (SIPO), China's supreme IP administration organ under the State Council.

The patent description determines whether the applicant can get the patent, while the claim determines the scope of the protection. The claim and description are the two most important documents in the process. As application documents are subject to publication,[11] there are usually concerns regarding IP infringement and the loss of core knowledge. Hence, many companies are reluctant to disclose the most sensitive parts of the technology. But limited disclosures often lead to limited protection, or even the failure of an application. Therefore, drafting the application in concise language is critical and complex: technical and IP specialists should work closely together on this. As the interviewed experts indicated, a simple translation of the existing patent into Chinese is not sufficient to provide comprehensive protection. Experienced local IP attorneys are more familiar with the specific wording in Chinese that is critical for ensuring protection without any loopholes. However, all the decisions made by the SIPO and Patent Re-examination Board (PRB) are subject to juridical review. If an application is rejected by the SIPO, a company can request a re-examination from the PRB. If the PRB rejects the application again, the company still has the possibility to appeal to the First People's Court for administrative litigation in Beijing.

Copyright of computer software

According to the Chinese Copyright Law, a copyright object needs to fulfil the following criteria: it must be in the literary, artistic or scientific domain; it must be original; and it must be able to be fixed in a tangible medium from which it can be perceived, reproduced or communicated (Implementing Regulation, Article 2). The protection term is the lifetime and 50 years after the author's death for natural persons (Copyright Law, Article 21). In the case of a legal entity or other organisations, the protection term is limited to 50 years. Since 2001 the current legal practice has extended protectable copyrights to include topics such as speeches and computer software (Article 3).

In 2002 China endorsed the Regulations for Protection of Computer Software. Protectable software must be developed independently by the developer and fixed on a tangible medium (Article 4). In addition, the copyright protection does not include the ideas, processing, operating methods, mathematical concepts or anything similar that are used in software development (Article 6). The protection term for software is the same as for usual copyright (Article 14). In general, copyright is not subject to registration, but the copyright registration of software is prima facie evidence of copyright ownership in case of a dispute. Although the registration of software requires the disclosure of source codes, the software owner has the possibility to conceal important contents of the source code, as the total registration should not exceed 60 pages (Chen, 2005).

In general, the SIPO does not as such consider software as patentable. Nevertheless, 'a software-related invention is patentable if it resolves a technical problem, uses a technical measure, and reaches a technical result' (Deng, 2004: 3). A business method can also be patented if it is a software program associated with a technical solution (Chen, 2005). Software may enjoy dual protection through both the Copyright Law and the Patent Law. A prerequisite for successful application for a software-related invention is a team of specialists with substantive experience.

Trade secrets

In China, 'trade secrets' must fulfil four criteria: they are business or technological information or know-how; they have economic value; they are not known to the public; and the owner has taken measures to keep them confidential.[12] However, the term 'public' is not precisely clarified in the law. According to interviewed IP attorneys, in legal practice the 'public' is considered a third party who has not signed a non-disclosure agreement (NDA) with the trade secret's possessor. Theoretically, almost everything can be proclaimed and protected as a trade secret without any applications. The protection term is infinite, as long as all the criteria are fulfilled.

Still, in practice, high-tech companies should on the whole not protect their technology in the form of trade secrets, especially not in areas where the technological life cycle is short. A fictional case explains why. A company generates a new type of technological knowledge, and keeps it as a trade secret. However, six months later a competitor introduces a product to the market with almost the same technology, and even applies for a patent. The original trade secret is no longer valid, as it is now known to the public. There is also the possibility that the competitor could have invented the technology using similar logical thinking and state-of-the-art R&D. According to several interviewees, this kind of scenario is not untypical as many high-tech companies are working in the same direction with similar ideas and R&D approaches. Even if the competitor has used illegal means to obtain this technology, the original company has been damaged and sometimes it is difficult to gather conclusive evidence. The protection effectiveness regarding core technologies' trade secrets is therefore limited. It is more appropriate to seal information regarding a specific R&D approach or the progress status of sensitive R&D projects as trade secrets, as there is no other means of protection.

Trademarks

According to the Trademark Law of China, protectable marks must be visible, and include any combination of words, designs, letters, numbers and three-dimensional marks. One principal criterion is that trademarks must clearly distinguish commodities or services from others (Article 8). The risk of trademark infringements in the form of counterfeit products could be high for high-tech companies, e.g. pharmaceutical companies such as Novartis or Roche, if they produce consumer products. However, if the high-tech company operates in the business-to-business area, the risk of trademark infringements is relatively low, since the players are well known and trademark infringements are easy to detect. Trademark applicants should first file the enquiry by submitting statutory application documents. Notably, each application can only cover one trademark in one class. If the application meets all the principal criteria, the trademark will be published and will become a registered trademark if in the following three months no opposition is filed. Each registration is valid for ten years and can be renewed indefinitely at ten-year intervals (Articles 20 and 38).

Industry-specific exclusive rights – IC layout design

In 2001 China endorsed the Regulation on Protection of IC Layout Designs. ICs or integrated circuits are fundamental components of a vast range of technology-intensive industries, such as computers and telecommunications. Granting exclusive rights to IC design owners not only shows China's commitment to its WTO accession, but was also done for the sake of its domestic semiconductor and telecommunication industries' development and to encourage foreign investment in these fields. In terms protection effectiveness, the recognition of an IC layout design's exclusive right is not as strong as an invention patent. This leads to the recommendation that both patent registration and the recognition of an IC layout design's exclusive right should be used, while a patent is always the best choice for the protection of core technology.

IPR administration bodies

In this section, the four IPR administration organisations in China will be briefly introduced. Each has its own specific administrative area and is responsible for applications for and enforcement of different types of IP.

State Intellectual Property Office (SIPO)

SIPO is the IP administration body responsible for patent applications and invalidation. Patent-related decisions made by SIPO and its Patent Re-examination Board (PRB)[13] are subject to reviews by the People's Court. If a patent applicant does not agree with the SIPO's decision, the PRB will be called into action. After another rejection by the PRB, the applicant still has a chance to appeal to the First Intermediate People's Court in Beijing for administrative litigation (Clark, 2000). Regional patent bureaux, set up by local governments, are responsible for local patent administration affairs but do not have the power to grant patent rights. They are not administrative organs of the SIPO. As a policy coordinator, the SIPO offers guidance on policy matters rather than giving concrete advice on specific cases. It is worth mentioning that the SIPO does not have the power to enforce IPRs, since it does not handle any patent disputes.

State Administration of Industry and Commerce (SAIC)

The SAIC is the governmental market supervisory body and the most powerful IP administrative organ in China. Its nationwide network consists of bureaux of four top-down administrative levels, and it exercises considerable and systematic supervision regarding law enforcement. In general, the SAIC is able to investigate IP infringement cases and has the most effective enforcement power against counterfeit products. For example, if an infringement is determined, the local SAIC bureau is able to stop the sale of the infringing items, order the destruction of the infringing products, impose fines and remove the machines used to produce the counterfeit goods (Chen, 2005). In order to track the increasing 'local protectionism', the local bureaux are required to report vertically to superiors at the next level until this reaches the SAIC in Beijing (Feng, 2003).

National Copyright Administration (NCA)

The National Copyright Administration is responsible for copyright administration and enforcement nationwide, including investigating infringement cases, administering foreign-related copyright issues, developing foreign-related arbitration rules and supervising the administrative authorities. The regional bureaux are a 'functional department' of local government, but they accept 'professional guidance' from the NCA.

General Administration of Customs (GAC)

Goods that infringe IPRs are not allowed to be imported or exported. The General Administration of Customs bans the import/export of IPR-infringing goods. Protectable IP includes patents, trademarks and copyright, and excludes the rights stipulated by the Law Against Unfair Competition. In order for the GAC to exercise this right, IP holders are required to record their IPRs with the GAC.[14] Only after an IP has been recorded can an IP holder who suspects that infringing goods are about to enter or exit China submit a written application to the GAC in which the necessary information is provided, such as the suspected infringer's name, specifications of the infringing goods, details of the consignment and evidence of infringement. When the GAC's investigation reveals infringement, it has the authority to confiscate, destroy or remove the goods and impose a fine (Feng, 2003).

IPR dispute settlement

The majority of IPR disputes remain civil disputes in China. There are four different ways to solve IPR disputes, and the involved parties are to a certain degree free to choose the way that they would like to settle the conflict. Each of the methods is appropriate for solving a specific state of affairs in an IP dispute. Thus knowing the pros and cons of the different methods is critical to achieving the most efficient and effective IPR dispute settlement.

Conciliation

Conciliation is probably the most 'peaceful' way to settle an IP dispute. The IPR holder can choose this way if the infringement only touches non-core IP and the damages or financial losses are not serious. If the dispute can be solved by conciliation, the IPR holder may ideally gain a new licence contract, or even build up future cooperation. However, there are still several uncertainties. First, it is important that the IPR holder discerns whether the infringer is honestly willing to solve this dispute peacefully, since the infringer may deliberately use conciliation to gain time. Second, the implementation of a conciliation agreement cannot be enforced by a court, which means that the infringer can change his or her mind at any time. If this happens, all the efforts made in the conciliation process will have been wasted resources. Therefore, as the interviewed experts agreed, conciliation is often not appropriate for high-tech industries, as the life cycles of many technologies are short and conciliation could be very inefficient and cause an unnecessary time-lag.

Administrative enforcement

Administrative enforcement will be carried out by government departments such as the SIPO, SAIC, NCA and their local bureaux, depending on the type of IP infringements found, e.g. trademark or copyright infringement. These departments have the authority to conduct administrative investigations that include evidence collection, confiscation of infringing products, imposition of fines and administrative conciliation (Li, 2005: 3). Dispute settlement through administrative enforcement is especially powerful and efficient. Generally, all the actions can be undertaken within a week. Compared with litigation, administrative enforcement is less costly and saves time, but it lacks the possibility of civil compensation (Chen, 2005). Most disputes encountered by foreign R&D are patent infringements, and the applicability of administrative enforcement in patent disputes is limited because the enforcement power of the SIPO, the supreme patent office in China, is relatively weak (see above).

Arbitration

According to the Arbitration Law, equal civil subjects may resolve their contractual disputes on property rights and interests through arbitration. Regional arbitration committees, such as that in Shanghai, and international ones, like the China International Economic and Trade Arbitration Committee (CIETAC), are authorised to accept claims regarding IP among equal natural persons, legal entities and other organisations (Article 2). If one party is a foreigner or foreign company, the dispute must be arbitrated by the CIETAC, which is located in Beijing and has two branches in Shanghai and Shenzhen. Compared with litigation, arbitration saves costs and time because the procedure is less complex. As an arbitration verdict can be compulsorily enforced by a court, the implementation is more effective than administrative enforcement. However, only when an arbitration agreement or clause has been inserted can the dispute be resolved in this way (Artlcle 4). Licence agreements or the results of joint R&D activities are often subject to IP-related disputes: arbitration is an appropriate way to solve these kinds of disputes if the contracts signed include an arbitration clause stating that arbitration committees have the authority to solve possible disputes.

Litigation

Litigation is the last means through which an IP dispute can be settled. All intermediate and high courts in China have benches that are in charge of IP hearings and the trials of IP-related cases. If a party is called to account for infringement, the court will, through civil prosecution, 'stop infringement, compensate any damages resulting from the infringement, make public apology to the injured party and reimburse the legal costs' (Li, 2005: 3). The winning party can request damages for either the lost profit or the infringer's illegal revenue. If the lost profit and illegal revenue cannot be proved, statutory damages of up to RMB500,000 may be awarded. The other litigation form is criminal prosecution. Since November 2004, the court may sentence the natural person or legal entity infringer to a set term of imprisonment and/or impose a fine, provided that the infringement satisfies the threshold required for a jail sentence.[15]

The IP strategy of an IPR owner determines which type of litigation should be considered. In general, the infringed party can execute both civil and criminal prosecutions. Through civil litigation, the infringed party is able to stop the infringing action and require indemnity. Successful criminal prosecution leads to higher public attention and has a strong deterring impact. In cases of serious IP infringements, companies can report to the Economic Crimes Investigation Division (ECID) of the Public Security Bureau (PSB), which could initiate the quick start of criminal prosecution (Neumann, 2005). Criminal prosecution gives a clear message on behalf of the company that it is very serious about any kind of IP infringement and will use any means to prevent and counter it. Nevertheless, successful criminal prosecutions of IP infringement in high-tech areas are still rare.

Comparison of different settlement methods

In a case of an IP infringement or dispute, the IPR method of settlement that owners should implement requires cautious consideration. Table 7.4 contrasts the benefits and risks of the four IP dispute settlement methods. The method chosen should depend on the infringement's context as well as on the MNC's general IP strategy in China. Different methods can also be adopted simultaneously. In general, legal action is rated as the best means of action against IP infringements by the respondents to the Swiss China Survey (CH-ina, 2005b). The most effective way to stop counterfeit products is administrative enforcement through the SAIC and NCA. In technology-intensive sectors, conciliation and litigation are more appropriate methods. Because litigation has the highest enforcement competence, it could be the most effective method against IP infringement; however, it is not the best way in every situation.

This case emerged through the interviews. Foreign high-tech company A has discovered that B, one of its major Chinese customers (a big SOE), has registered a patent in China that is an exact copy of one of A's patents in Europe. This patent is also used in the products sold to B. After a thorough assessment, the company decided not to litigate against B but to solve the problem through conciliation. In the end, A was able to obtain additional orders from its Chinese customer as a kind of 'compensation'. Company A's decision-makers did not see litigation as the most effective way to handle the situation because, first, the litigation process is usually very time-consuming and absorbs tremendous resources. Second, as B is a large and influential SOE, the chances of winning the case were not high due to local protectionism. Third, even if A could win the lawsuit, this gain would hardly compensate for its losses in terms of

Table 7.4 Balancing different dispute settlement methods

Action	Benefit	Risk
Conciliation	Little harm to relationship Possible new gain from outdated technology through new licence	Uncertainty about counterpart's sincerity No enforcement power Risk of time and resource waste
Administrative enforcement	Lower cost and informal procedure High degree of publicity discourages potential future infringements Chance to build relationship with local administration	Compensation is low or even non-existent IPR administration cannot take compulsory measures Regional protectionism and corruption could be a barrier
Arbitration	Efficient and effective Low cost and less complex procedure Appropriate for multiple-jurisdiction disputes Confidentiality Verdict implemented by court	No arbitration clause, no arbitration Patent invalidation is not subject to arbitration Third-party involvement is not subject to arbitration
Civil litigation	Preliminary injunction possible Strong verdict implementation Civil compensation possible	Time- and resource-consuming Risk of losing lawsuit
Criminal prosecution	High degree of publicity discourages potential future infringements	Conviction unlikely due to high standard of evidence required

Source: Based on Feng (2003) and James (2004)

an important customer and its harmonious relationship with the local authority. One of the most important issues was perhaps A's assessment of the infringed technology's real value and strategic importance. If the infringed technology had been of core value and high strategic importance to A, conciliation may not have been the appropriate approach to settle this dispute. Yet conciliation cannot fully protect IPRs as it cannot be enforced. The implication drawn from this case is that when choosing settlement methods for IPR disputes, decision-makers should look at the state of affairs from a more holistic perspective. 'We are in the right' should not be the only motivation to start a litigation process.

This section has explained the current IPR protection system in China, and discussed protectable IPRs and their applicability, and key IP administrative organs and their responsibilities. Different types of IP dispute settlement have also been explained. The next section delves deeper into managing IPRs in practice, revealing the major uncertainties faced by foreign R&D centres in China.

Uncertainties in IP management practice in China

Being familiar with the mechanisms of the IPR protection system will not be sufficient for foreign R&D to manage IPRs properly. More important is awareness of the ambiguities and pitfalls of managing IP. This section discusses the IP-related uncertainties that MNCs have to face from two different perspectives: environmental and managerial (see Table 7.5). This is followed by discussion of the managerial implications identified and gathered through the research. Environmental uncertainties are beyond the R&D organisation's control, but awareness could help MNCs to be prepared for certain complex situations and fine-tune their IP strategy accordingly. Managerial uncertainty is an area where decision-makers are able take

Table 7.5 **Areas of major uncertainties in IP management practice in China**

Environmental uncertainties	Managerial uncertainties
IP legal system	IP ownership
IP legislation (inconsistency with TRIPS agreement)	IP leakage
IP administrative system	IP ownership dispute
	Corporate IP-related coordination
	Cross-border IP transfer

more control. The awareness of the R&D manager should result in concrete management actions, thus keeping uncertainties at the lowest level.

Environmental uncertainties

IP legal system

Notably, China's government has devoted huge efforts to establishing a more transparent and efficient IP legal system in recent years. However, compared with international benchmarks, the country's IP legal system reveals particular weaknesses in litigation and enforcement. The field research has identified several weaknesses in these areas that are the major sources of environmental uncertainties. Based on the interview results and analysis, the managerial implications of encountering uncertainties in the IP legal system will be provided.

Low effectiveness of litigation

The litigation process in China is often less transparent for outsiders and lacks effectiveness. The following case[16] depicts typical situations encountered by domestic and foreign companies during IPR-related litigation. A, a Shanghai-based computer manufacturer, sued B, a medium-sized computer manufacturer in city X, for violating its utility model patent. City X, which is a provincial capital, was selected as the place where the trial will occur. As soon as B received notice that charges were being brought against it, it immediately submitted an invalidation request of the patent to the PRB. This is a typical tactical move by a defendant in an IP dispute, since the court is by law authorised to suspend a litigation process until the PRB has made its final decision. Through this process, the defendant gained more time to prepare for the litigation. Although the PRB rejected B's claim after examination, the court did not continue the trial because the local judges were not familiar with the specific technology and external experts were required. A was obliged to pay the expert fee first. Several months later, the court continued the trial. As it was quite obvious that B had infringed A's utility model patent, the trial quickly turned to the question of the amount of damages. According to China's Patent Law (Article 60), one party can request damages in respect of either a loss of profit or the infringer's illegal revenue. If the lost profit and illegal revenue cannot be exactly proven, statutory damages may be awarded to a maximum of RMB500,000 or US$60,000.[17] In this case, the lost profit and the revenue could not be clearly quantified and the court determined the amount of damages according to its estimation. The final verdict was reached three months later. A won the lawsuit, and B had to make a public apology and pay RMB50,000 in compensation. The amount of compensation was definitively lower than the real damage to A. A's request for a retrial was rejected by the higher court. The whole trial took more then two-and-a-half years.

As revealed in this case, plaintiffs often struggle through the IP litigation and have a chance to win if the evidence is there. But the effectiveness of the law is limited due to the lengthy litigation process and the court's deficient determining of penalties. A long-lasting IP dispute process is definitively to the plaintiff's disadvantage, hence the defendant's strategy is often to play for time. It is even likely that by the time the sentence has been determined by the court, the technology under dispute is already outdated due to technology's rapid development. A further disadvantage for the infringed party is that if the damage cannot be exactly proven, which is quite usual, US$60,000 is the maximum statutory compensation that can be set by law. Since China still considers itself as a developing country, and has huge economic disparities, the maximum amount of statutory compensation is set very conservatively by the authorities. Fines and criminal prosecution have not been established by the Chinese legal system as regular means of punishment. In general, compensation ordered by the courts is often more symbolic. As one interviewee pointed out, 'the cost of IP infringement in China is far too low to deter other IP infringers'. Hence, a realistic goal for any litigation in China is to stop further infringement.

The reasons behind China's low litigation effectiveness are multifaceted. Some IPR dispute cases overburden the local courts' competence. Although the IPR judges' professionalism has steadily improved and the government now provides special training for judges, there is a clear shortage of adequately trained judges. The judiciary's IPR-related expertise is especially insufficient outside the major cities. For example, in the most advanced economic regions in China, Shanghai and the nearby Jiangsu province, the high and intermediate courts' IP-related judgment body has only been established for ten years. At present only 40 per cent of the high and intermediate courts' IPR judges have a master's degree (Judicial Protection of IPR in China, 2005). In addition, ambiguous statutes, complicated administrative procedures and legislative loopholes, such as the punishment of IP infringement, are further causes of deficiencies at the institutional level. Local protectionism, corruption and bureaucracy have further diminished IP litigation's transparency and effectiveness and nullified the advancements made.

Weak enforcement

Besides litigation's low effectiveness, China's enforcement of IPRs is still weak. IPR infringers are often not prosecuted or punished effectively. The implementation of verdicts may only occur after long delays. Above all, there are difficulties in enforcing a verdict in other cities or remote regions. As mentioned before, local protectionism is definitely one of the major reasons for China's weak enforcement. If an infringing enterprise is an important employer and tax revenue contributor, it often enjoys strong backing from the local government. It is worth mentioning that company B, the infringer in the previous case, is one of the major employers in city X and has the local government as one of its stakeholders. In order to protect the local economy, local officials may administratively hinder a judicial trial and be reluctant to implement a verdict, or even impede implementation. In specific cases, enforcement is also weakened by corruption. Raids or close-down actions could, for example, be delayed, or an infringer could be informed in advance so that it has enough time to move out of its facilities.

Another barrier to enforcement is the vertical or horizontal joint coordination in administrative bodies.

While a verdict can be implemented comparatively smoothly in developed regions such as Beijing, Shanghai, Jiangsu and Zhejiang, delays in implementation are common in more remote areas of the country. Bureaucracy and a lack of qualified personnel in local administrations are also responsible for the weak enforcement of the IP legal system. It is worth mentioning that currently there are companies in China specialising in IP infringement detection and IPR protection enforcement. Mainly founded by former state officials and employing a mixture of private detectives and legal counsellors, they often have good relationship networks and many years of expertise in the area. These types of companies are frequently very efficient and could be regarded as accelerators of IPR enforcement.

Managerial recommendations

The discussion above has revealed that, under certain circumstances, conciliation could be a good way of solving IP disputes. However, as mentioned previously, in many cases litigation remains the most effective method against IP infringement in terms of enforcement competence. Once a company has decided to litigate against an IP infringer in China, it will be crucial to pay attention to a number of issues in order to increase the effectiveness of the legal action.

First, as agreed by most of the interviewed executives and IP attorneys, a very useful means of achieving effective litigation and enforcement is to gain public attention. As discussed above, the central government is keen to enhance IPR protection in the country and improve China's image in this area. Much public awareness will attract the attention of high-level officials and put pressure on the local authority. At the same time, a PR campaign could convey the company's seriousness in combating any kind of IP infringement. Making clear that the company will pursue every single case of IP infringement at any costs is an effective method to deter further infringement. Furthermore, in China there is an invisible penalty for infringers: the loss of 'face' when losing a lawsuit. PR is especially effective in industry sectors in which all the players are well known.

Second, choosing the most favourable and competent court with regard to the IP infringement is also key in enhancing effectiveness. By law, all local courts can judge an infringement case when an IP-infringement-related activity occurs within the geographical area of the court's jurisdiction. However, an IPR holder should choose the most advantageous place for the trial, i.e. where there is a court and judges with the most experience in IP-related disputes (Li, 2005). Not only could this decrease the probability of local protectionism and corruption, but it could achieve a better litigation result. Generally, Beijing and Shanghai courts are preferred above others. Another approach that large MNCs often use is to sue the infringer overseas if the infringing party has an overseas subsidiary and is selling the product abroad (e.g. Cisco sued Huawei in the USA). The difficulty is that a sentence delivered abroad has little chance of being enforced in China.

Third, hiring qualified IP lawyers with substantial experience is a crucial step to avoid uncertainties during IP litigation. The right IP attorney will not only help to win the case, but could assist with successfully enforcing the verdict. Well-known law firms often maintain good relationships with legal authorities. Some of these law firms even have a state-owned background, and some of the lawyers and senior partners are former judges and officials in the Chinese legal system. As mentioned by one experienced interviewee, there are also a few private companies, such as the China

United Intellectual Property Protection Centre, specialising in IP infringement investigation and with branches in various major cities. These types of 'service providers' do not only resolve IP infringement cases, but are also able to carry out effective enforcement due to their good relations with different levels of government. In fact, these private companies are considered semi-governmental, since their founders are usually former high-ranking governmental officials working in public security or as legal authorities. Many well-known MNCs are among their customers. Furthermore, becoming a member of IPR organisations with a strong enforcement network, such as the Quality Brands Protection Committee, could also increase the effectiveness of enforcement (Neumann, 2005).

If the IPR holder encounters strong resistance during the litigation or enforcement process due to local protectionism, a last step could be to appeal to the State Office of Intellectual Property Protection. Initiated by the central government, the office is a national working group dedicated to IPR protection and aims to strengthen IP protection further in the country. Led by the vice premier, the working group consists of 13 members who are in charge of 12 IPR-related agencies (State Intellectual Property Office, 2005b). The interviewed attorneys consider appealing to the State Office of Intellectual Property Protection as a final effective approach to solve a delicate IP dispute. However, strong evidence is a prerequisite to attract this body's attention.

Consistency with TRIPS agreement

Although China's IP legislation is very close to the TRIPS agreement standard, inconsistencies still remain due to the country's indigenous perspective and perhaps for reasons of policy. For instance, in terms of patents, almost all inventions should be patentable according to the TRIPS standard. But Article 5 of China's Patent Law stipulates that patentable inventions must meet three criteria: the invention does not violate the laws of China; it does not contradict social morality; and it does not prejudice the public interest. These requirements seem reasonable to Chinese legislative authorities, but some foreign researchers believe that this general definition is overly broad and leaves too much room for interpretation, thus justifying the exclusion of, or having protection withdrawn from, certain inventions (Moga and Jonathan, 2002). Another example is that Chinese IP law does not clearly indicate that products under compulsory licence should mainly supply the domestic market, as required by the TRIPS agreement. Furthermore, while the TRIPS agreement requires adequate remuneration of the licensor (Article 31(h)), China's Patent Law (Article 54) empowers the patent administrative body under the State Council to adjudicate if the licensor and licensee fail to agree on an 'exploitation fee' (Moga and Jonathan, 2002). 'These differences may prove to be significant in the future, but so far no single case of compulsory use of a patent has occurred since the establishment of the patent regime in China' (Weinstein and Fernandez, 2004: 232). With regard to the copyright of computer software, Chinese regulation, in contrast to the TRIPS agreement, does allow any use of software for training purposes,[18] which implicitly allows reproduction beyond the licence.

China's compliance with the TRIPS agreement in the course of its WTO accession does not necessarily mean that the country's IP legislation as yet equates to the TRIPS standard. In particular, broad definitions or indigenous interpretations of certain TRIPS standards force MNCs to pay extra attention to those inconsistencies that could lead to unwelcome surprises in cases of dispute. However, China is not the only WTO member country whose IP law deviates somewhat from the TRIPS agreement.

IP administrative system

In line with the findings from different literature sources (e.g. Yang, 2003b; Feng, 2003; James, 2004), interviewed R&D managers and IP attorneys pointed out that China's administrative system still has quite a few acute weaknesses. For example, one interviewed company reported that the patent authority had granted one of its patents to another party. There could be several reasons for this type of error. First, there are deficits in the archival system's hardware and software. Patent literature archival systems reveal weaknesses in terms of comprehensiveness, stability, capacity and currency (Tang and Tian, 2005). Second, despite continuing improvements, there is a lack of well-trained and experienced patent examiners. China is seeing an explosion of patent applications: since the beginning of the new millennium, patent applications have increased more than 20 per cent each year (*China IPR News*, 2005). Between 2001 and 2004 more than 1 million patent applications were received by the SIPO. The IP administration has speeded up the education and hiring of patent examiners, but the country cannot educate examiners fast enough to keep track of the number of applications. By the end of 2004 there were around 1,300 professional examiners working on the examination of patents, 3.7 times as many as in 1995 (*China IPR News*, 2005). The SIPO aims to recruit around 500 new patent examiners each year until 2010 (*HR Weekly*, 2004). It might be easy to increase the quantity of patent examiners within a short period, but it is very difficult and will take much longer to build up a well-educated and experienced generation of examiners.

Managerial uncertainties

Companies are almost unable to prevent totally any type of IP infringement and dispute. The developing character of the Chinese IP protection environment causes uncertainties that a company can barely control. However, the risk of IP infringements and disputes can be diminished if a foreign R&D centre is able to take control of the controllable. The following sections discuss five identified risk-bearing, but controllable, domains of IP management practice in China. Awareness and successful management of these uncertainties enable international R&D organisations to prevail and avoid many of the risks associated with IP in China.

IP ownership

One of the first steps in ensuring IP protection is to make sure that the company has a solid IP ownership in the country. First, MNCs need to build the patent's legal foundation in China in order to make it protectable, as patent protection has regional limitations. A filed patent also needs to be solid enough. In other words, MNCs need to decrease their patents' vulnerability to invalidation claims.

Regional limitation of patent protection

Field research has revealed that some MNCs believe the most secure way to protect their IP is not to bring the technology into the country. This prevailing attitude is confirmed by the Swiss China Survey, according to which refraining from applying advanced technologies in China is the second most promising measure to prevent IP infringement, after legal action (CH-ina, 2005b). Foreign companies are worried about potential IP infringements through patent publication and are thus reluctant to file their existing patents in China. Considering the developing character of the Chinese IP protection system, this kind of thinking seems to be reasonable. However, in practice this IP protection strategy also has high risks, especially when the technology is and will be used in the Chinese market. The following case represents the type of IP infringement/dispute that MNCs may encounter.[19] A European mechanical engineering manufacturer E is in possession of a patent filed in Europe. This patent is used in a product that is introduced to the Chinese market. However, after launching the product in China, company E was sued by a local competitor F for infringing its patent. It turned out that company F held the patent right in China. After years of litigation, E lost the lawsuit and even had to pay F compensation. Different reasons, such as an insufficient examination of the patent's 'novelty' by the Chinese patent authority, local protectionism or even coincidence, made it possible for F to be the patent's IPR owner in China, even though F's patent is very similar to E's patent in Europe.

Valuable implications can be drawn from this case, regardless of which reason is the more relevant. R&D managers should realise that foreign-filed patents are not protected by China's Patent Law. In China, IP only enjoys legal protection if a patent is filed locally or via a Patent Cooperation Treaty (PCT) application that allows applicants to file the same patent in different member countries simultaneously. As long as IP has no valid legal foundation, people can copy and use it without fear of retribution. Due to the remaining uncertainties in the IP administrative system, MNCs should invest even more effort in proactively securing their IP locally by filing local patents. This is especially relevant in the area of R&D.

Vulnerability of patents

Possession of solid IP ownership also depends on a patent's robustness. According to estimations, about 30 per cent of patents are invalidated in China each year (Chen, 2005). In the pharmaceutical industry, two prominent examples have specifically shown how important it is to have a strong patent.

In 1994 the US pharmaceutical company Pfizer filed a patent application for sildenafil citrate[20] for medicinal use in China. After seven years of examination, the SIPO granted the patent. Nevertheless, 12 domestic medicine companies challenged the patent's validity. Finally, the PRB invalidated the patent on the ground that it had been insufficiently disclosed that the claimed invention violated Article 26 of the Patent Law.[21] In other words, the patent's manual had failed to provide convincing technical content (*People's Daily*, 2004c). A similar experience was had by the British pharmaceutical company GlaxoSmithKline (GSK). Several Chinese pharmaceutical companies filed petitions to invalidate GSK's patent on rosiglitazone, one of the three ingredients of Avandi, a popular anti-diabetic drug on the international market. The claims stated that the invention was not novel and had already been made public before the patent was filed in China (Judicial Protection of IPR in China, 2005). If the claims could be proven, the patent would be invalidated. However, at the beginning of the oral hearing GSK voluntarily abandoned its patent (Bai and Cheng, 2004); in return, it succeeded in having the invalidation claims withdrawn. This settlement allowed the Chinese companies to proceed with patenting their own versions of the drug in China and GSK escaped the risk that the court could declare its patent invalid from the date it was issued (Chen, 2005).

The challenged patents in both the above-mentioned cases are keys to a huge Chinese market. The more lucrative a patent is, the more likely it is to be challenged by local competitors. Although the decisions of the Chinese IP and legal authorities are controversial and certain foreign experts believe that political influence was exerted, a robust patent is definitively the keystone of IP protection, and is particularly important within the Chinese environment. Some foreign companies tend to withhold certain key knowledge during the publication of a patent in China in order to protect this knowledge. Even though such a patent application could pass the examination, the patent has to withstand much greater scrutiny if there is an invalidation claim. Thus R&D managers should cautiously balance the risk of patent invalidation and the benefits received from restricted disclosure.

IP leakages

IP can be protected more effectively if all leakages can be prevented. It is instrumental for MNCs to build up a clear understanding of potential leakages. In general there are two kinds of IP leakages: internal and external.

Internal IP leakage

Internal IP leakages are caused by company employees. All the interviewed international R&D organisations have technical, contractual and educational means by which they aim to prevent such leakage. Technical security measures include firewalls and forbidding the use of private storage media such as portable hard disks or USB memory sticks. However, due to the technology's high sensitivity and industry-specific needs, one of the interviewed R&D centres has set up a comprehensive e-mail filter system and no use of the internet is allowed at work. Contractual means of prevention include the signing of a non-disclosure agreement (NDA)[22] and/or a non-competition agreement (NCA). An NDA is viewed as an employee's obligation not to disclose information covered by the agreement. If there is a dispute regarding the leakage of trade secrets, this agreement acts as proof that the employer took measures to keep the information confidential. An NCA prevents the employee from competing with the employer by limiting him/her from working in the same field after leaving his/her employment. Although there is at present no real judicial regulation regarding an employer's duty to compensate an employee once an NCA has been signed, the interviewed attorneys maintain that, legally, the NCA would be invalid if it does not include a clause on compensating the employee for limiting his/her freedom to choose a job. Notably, with the exception of core technical staff, there is no need to have all employees sign an NCA. As mentioned by one interviewed manager, maintaining good relationships with and keeping track of resigned key employees are effective ways to thwart IP leakage as well.

Another critical instrument to prevent internal IP leakage, besides technical and contractual means, is IP education. As mentioned previously, there is still a lack of profound social awareness of IP in China and a prevalent knowledge gap about the subject. Consequently, comprehensive IP-related education of R&D staff is a crucial part of job training. One of the interviewees articulated it thus: 'IP education should not only explain the basics and its importance for the company. Most importantly, it has to clarify the importance of IP protection for each individual R&D staff member as part of the company.'

The mixture of technical, contractual and educational means could effectively prevent internal IP leakage to a certain degree, and they are easy to implement. However, in reality employers are hardly able to control what their employees do. Hence, the best way to achieve sustainable prevention of internal IP leakage is to win the hearts and loyalty of employees, but this might also be the most challenging way.

External IP leakages

IP leakages can occur through external partners such as suppliers, customers and third parties. While outsourcing has become a growing trend for companies in the triad regions in order to reduce costs and reach a higher degree of specialisation, in China many companies still have concerns about extending the supplier base. Several interviewed companies say they have even reduced their number of suppliers in order to reduce the risk of IP leakage. Certain large state-owned companies are often stakeholders in different subsidiaries engaged in a similar or related business area. The likelihood of unwanted technology transfer increases when partnering a local company involved in a large and less transparent network.

Another form of external leakage is more difficult to handle: through state agencies such as certification organisations. As in all markets with regulations on industrial norms, companies in China have to pass quality tests and gain certification for their products in order to obtain accreditation for market access. Agencies issuing these types of certificates are appointed by the specific industry's central department. However, local agencies, especially in less developed regions, often reveal huge internal IP loopholes due to non-transparent relationship networks and even corruption. The worst scenario is for a company to see its blueprints being circulated by local companies, and at least two of the interviewed firms have had such an experience. So far, this kind of IP leakage and infringement has been limited to individual cases. The implication that can be drawn is that each external partner with access to a company's IP needs to be carefully monitored and evaluated in terms of background and its IP infringement track record.

IP leakages during patent application

As revealed earlier, the examination procedures of invention and utility model patents differ. An invention patent application also has uncertainties and certain risk. After an invention patent's preliminary examination, all the application documents will be published. Other parties could use this publicly accessible information to produce products or to apply for a utility model patent. Since a utility model patent is not subject to substantive examination and can be obtained much faster, it is possible that other parties could obtain a patent before the original invention patent has been granted. This is regarded as a new kind of 'IP piracy' (Chen, 2005). Although the invention patent applicant can request the PRB to annul the infringing utility model, the process will require extra resources from the patent owner. A managerial implication suggested by several of the interviewed IP attorneys is to apply for both an invention and a utility model patent simultaneously during the invention process. The utility model patent can then be used as an effective means of protection during the application procedure, while an invention patent still undoubtedly offers the best protection once the invention has been completed.

IP ownership disputes

IP ownership disputes discussed here are caused less by intentional IP infringement and rather more by managerial ambiguities. The research has identified the two most important types of IP ownership disputes: service invention disputes between an employee and an employer (internal disputes), and IP ownership disputes between cooperation partners (external disputes). A focal point of IP ownership disputes is a lack of clarity about the owner of a specific IPR.

Service invention disputes (internal disputes)

Disputes between employees and employers regarding IP ownership are of central importance, particularly for R&D organisations. Often an employer believes that all IP outcome from an employee belongs to the firm and is stipulated thus in the employment contract. However, an employer's claim to his current or former employee's invention outcome will only be legally supported if the invention fulfils 'service invention' conditions, as they are termed; thus the controversial issue is whether an invention is a service invention. According to the Implementation Regulation of China's Patent Law (Article 11), a service invention is an invention that was developed in the course of an employee performing his duty and other tasks assigned to him; or if the invention relates to someone's duty or other task entrusted to him by a former employer within one year of his resignation, retirement or change of work; or if an employee uses a firm's materials and technological means to develop an invention that does not relate to his duty or task (Patent Law, Article 6). It is noteworthy that the definition of a 'service invention' in current Chinese IP law does not differ from that of international benchmark definitions. If there is litigation, the plaintiff, usually the company, has to provide evidence that the invention is indeed a service invention. Given that expertise and knowledge are the foundations of a new invention, service invention disputes usually involve experienced researchers. In general, the probability of such disputes decreases with the increased sophistication of the technology and equipment involved and a research object's life cycle. Invention patents in certain industries also require interdisciplinary knowledge and many years of investment. Thus a utility model patent, which is less sophisticated than an invention patent, is more frequently the subject of this sort of IP dispute.

Whether an invention was developed in the course of an employee merely performing his duty is often quite obvious. If specific tools and equipment were required for the invention process, it is also easy to provide evidence of whether the employee relied on his own equipment during the process.[23] A complex situation occurs if the invention is not clearly related to an R&D staff member's duty or task, or if the invention only required conventional equipment such as a normal PC. Besides hardware, inventions are heavily based on an employee's specific knowledge and experience. Many employers think that as soon as an invention involves an R&D staff member's duty-related specific knowledge, any intentional outcome from the employee is a service invention. However, according to the interviewed law experts, these beliefs can only be legally supported if the involved knowledge is a type of company trade secret. In practice, to what extent an R&D staff member's cumulative tacit knowledge and expertise can be declared a company trade secret, in other words property of the company, is often the nub of the argument.

The following case provided by the China Legal Information Centre (2005) illustrates these issues graphically. Mr H was a researcher at G Pharmacy, a chemical institute. In 1979 he was assigned to develop a method to extract a specific natural toxin. The project was completed in 1982. In 1985 Mr H was inspired by an article in an academic journal on the potential use of this toxin for drug-withdrawal symptoms. He decided to research this field privately in his own time and using his own equipment. After ten years of research, Mr H and his partner finally developed a product and successfully applied for two invention patents in 1994 and 1995. It is worth mentioning that he remained an employee of G Pharmacy during this period. In 1997 G Pharmacy and the Chemistry Institute filed a lawsuit claiming that these two inventions were service inventions. After years of litigation, Mr H won the lawsuit, as he was able to prove that the invention had been made in his own time and with his private equipment. Above all, although Mr H had done research on the same toxin as the one with which he had dealt during his project at work, the focuses of his work project and his later invention were totally different. His work task had been to focus on this toxin's extract technology, but his invention focused on the toxin's special utilisation, namely in drug-withdrawal symptoms. Inspiration for his invention had come from an academic journal article, a publicly accessible resource. Mr H had most certainly gained specific knowledge of this toxin through his previous duty, but the court obviously did not consider this knowledge to be his employer's property, or else G Pharmacy had simply failed to prove that the involved knowledge was a trade secret.

In a nutshell, in order to avoid uncertainties in dealing with internal IP disputes, R&D organisations need first to have a clear understanding of the legal connotations with regard to service inventions. Not only should management be familiar with this issue, but also every single R&D staff member needs to be aware of it. Furthermore, an employment contract sometimes stipulates that all IP outcomes from an R&D staff member belong to the firm, without specifying this further. This kind of generalisation could lead to the invalidation of the employment contract for deviating from the legal definition of a service invention. Thus, in order to avoid ambiguity, the employment contract should stick closely to the legal definition of a service invention.

IPR ownership disputes between cooperation partners (external disputes)

Besides IPR disputes caused by deliberate IP infringements, as discussed previously, external IPR ownership disputes can also occur with cooperation partners. As the IP issue is a central subject of cooperation negotiations, disputes are not usually caused by ambiguous agreements. More often, the IP of research by-products that were neglected or overlooked during the negotiation has a great potential to ignite disputes. Thus, R&D cooperation partners need to clarify the IP issue carefully beforehand and take research by-products into consideration. It is worth mentioning that many local R&D cooperation partners, such as Chinese universities, still have little experience with IP issues within R&D cooperation, thus patience is often required by foreign partners.

Corporate IP-related coordination

The interviewed IP lawyers stressed the importance of corporate coordination in managing IP. Insufficient communication, unclear responsibilities and a lack of coordination between engineers and IP attorneys will weaken a patent's robustness and also the chance of winning a lawsuit if there is an IP dispute. In the Chinese context in particular, one needs to take two further aspects into consideration when working with the local lawyers. First, the language is a challenging issue. As mentioned previously, merely translating the existing patent into Chinese is definitively not sufficient to secure a solid patent. Experienced local IP attorneys are more familiar with the fact that the indigenous wording is critical to ensure the entire scope of protection without any loopholes. Second, choosing the right legal representative is critical. At some well-known law firms a few of the lawyers have worked as judges, or had important positions within the legal system. The law firm's good relationship network could be very helpful. It is worth mentioning that a long-term relationship enables an external IP lawyer to build up a better understanding of the technology, and this will support smooth cooperation. But despite the rich experience an IP lawyer may have, in terms of the technical details scientists and engineers working in R&D will always be more familiar with the content of the invention. Hence, MNCs need to assign the most capable engineers and employ the most experienced local IP attorneys to handle the IP issue together.

Besides the R&D department, IPR protection should involve other departments such as marketing and distribution. In general, the sooner IP infringements can be tracked down, the sooner steps can be taken against the infringement and the better the extent of the damage can be minimised. R&D departments often do not have extensive direct interactions with markets, but do have the competency to detect IP infringement. Vice versa, marketing and distribution departments often have extensive knowledge about competitors' markets and products, but lack technical expertise to identify IP infringement. Particularly in high-tech areas, IP piracies are difficult to discover without an in-depth technological knowledge. Consequently, it is essential for a company to have a thorough structural integration of IP infringement monitoring and detection mechanisms that involve interdisciplinary parties, such as distribution, technical and legal departments, to protect its IPRs.

Cross-border IP transfer

As R&D internationalisation is developing rapidly, many MNCs have installed different R&D facilities around the globe. Concurrently, R&D has become more sophisticated, so that one invention requires varied competencies and may combine contributions from different, regionally distributed R&D centres. Hence, cross-border IP transfer has become critical for international R&D activities. However, the uncertainty is that, like almost every other country, China has certain restrictions regarding technology export.

According to the Regulation on Restriction of Technology Import and Export, technologies are divided into three categories: prohibited, restricted[24] and permitted. Only a small portion of technology is prohibited or restricted in respect of export. Although government approval is not required for permitted technology, a technology transaction nevertheless requires online registration on the Ministry of Commerce's website and royalty payments. Current IT systems enable knowledge and technology to be transferred around the globe in no time, and it is thus very difficult for the government to control this form of technology export. Thus, despite the regulations, not all technology transactions are registered by foreign companies. Nevertheless, the awareness of restrictions and export regulations can help foreign R&D to avoid unnecessary uncertainties and ensure sustainable development.

The transfer of patents or the application rights of a patent to a foreign entity is subject to official approval (Patent Law, Article 10). Due to the remaining mistrust of the IP protection system in China, some MNCs are reluctant to file a patent in the country first. One way to avoid uncertainties in this matter is to file the patent via a Patent Cooperation Treaty (PCT) application. The PCT allows an applicant to file the same application in different member countries simultaneously. It also grants the applicant priority rights, which means that the earliest filing date is viewed as each application's filing date in different countries, if these applications are filed within the priority period. The priority period of an invention is 12 months, while utility model and design patents have a six-month priority right.[25] The strategy of some MNCs is to obtain priority right in their country by filing a PCT application in China. Moreover, compared with other countries, the fee for the PCT application in China is very reasonable (Xu, 2004).

Conclusion

Less than 25 years ago, China started to build up an IP protection system from scratch. China's WTO accession and the country's ambitious goal to become a leading economy through advancement in science and technology have leveraged the fast development of the IP protection system. Due to multiple environmental and managerial uncertainties, discussed above, managing IPRs within the Chinese context remains a challenge for international companies and their R&D activities. Building an advanced IP protection system in China still requires much effort and time. Many underlying reasons, such as the weak social awareness of IP, lack of IP-related experience, insufficient IP personnel, loopholes in IP administrative and legislative systems, protectionism and corruption, are beyond MNCs' control. Yet field research has revealed that MNCs and R&D managers could overcome these barriers and diminish uncertainties by heeding and integrating the following implications.

Awareness of uncertainties

MNC executives and R&D managers need first to make themselves aware of the existing uncertainties. A thorough understanding of the Chinese IP environment is therefore critical. Although allowing experienced professionals to handle IPRs is necessary, this is not sufficient. Regular and coordinated communication between R&D managers and IP experts gives R&D organisations solid ground to walk on in terms of IP protection during the entire R&D process. Potential IP leakages should be detected early. MNCs should also keep track of the dynamic process of development regarding regulations and changes in legislation. Furthermore, an effective IP infringement monitoring mechanism is key to minimising damage through IP infringement.

Robustness of IP ownership

Protecting IP in a less mature environment requires higher IPR robustness to withstand different kinds of uncertainties. First, MNCs must have protectable IPRs. In other words, IPRs need to be built on a sound legal foundation, e.g. through the registration of patents. Second, IPRs must be solid, as this is crucial to fend off potential attacks. Solid IP ownership also means that there are no IP leakages. For the protection of IP such as trade secrets (e.g. knowledge of customers' specific needs) that cannot be filed or registered, MNCs need to enforce carefully all technical, contractual and educational means to prevent leakages. Usually people and not technical hardware are responsible for IP leakages; thus building loyalty with R&D staff will be key to prevent internal leakages. External partners need to be carefully selected and monitored. Spreading or holding back a core part of sensitive knowledge and technology is also considered an effective way to guard against IP leakages.

Relationship building

To a certain degree, deficits revealed in the IPR protection framework can be compensated by a healthy relationship with the authorities. It is less complicated to start a fruitful dialogue with the local government and IP administration agencies, especially in respect of foreign R&D activities. MNCs that have invested in R&D in China have shown their trust in the country and its future, which officials are glad to acknowledge. MNCs should take advantage of this official support and use each opportunity for dialogue to build a lasting relationship with the authorities. Personal interaction with the local officials enables MNCs to talk about their concerns and problems, and they are very likely to receive direct support. By being a part of the local community, MNCs are also expected to share their experiences to improve IPR protection further in China.

Notes

1 The Swiss China Survey is an in-depth survey conducted among Swiss companies active or interested in China in 2004–2005 (see CH-ina, 2005b).
2 As stated by the former director-general of the World Intellectual Property Organization, Dr Arpard Bogsch, the speed with which China achieved this development is unique in IP's world history (Li, 2005).
3 For example, one of China's goals is to become one of the ten most S&T-competitive nations by 2010.
4 Laws and regulations cited in the endnotes are based on Legislation Working Group of National People's Congress (2003).
5 Invention: new technical solution related to a product, a process or improvement of a product. An invention patent enjoys 20 years of protection (Implementing Regulation of the Patent Law, Article 2).
6 Utility model: new technical solution related to the shape, structure, or their combination, of a product fit for a practical use. A utility model patent enjoys ten years of protection (Implementing Regulation of the Patent Law, Article 2).
7 Design: new design of the shape, pattern, colour, or their combination, of a product which creates an aesthetic feeling and is fit for industrial application. A design patent enjoys ten years of protection (Implementing Regulation of the Patent Law, Article 2).
8 A request shall state the title of the invention or utility model, the name of the inventor or creator, the name and address of the applicant and related matters (Patent Law, Article 26).
9 A description shall be given of the invention or utility model in a manner sufficiently clear and complete so as to enable a

person skilled in the relevant field of technology to carry it out. If necessary, drawings are required (Patent Law, Article 26).

10 An abstract shall briefly state the main technical points of the invention or utility model (Patent Law, Article 26).

11 Only patent applications with national defence-related technology are not disclosed.

12 Law Against Unfair Competition, Article 10.

13 The Patent Re-examination Board is constituted by SIPO directors and assigned experts in relevant fields (Implementing Regulation of Patent Law, Article 58).

14 'The record is valid for seven years, extendable for another seven years, provided the relevant IPRs are valid' (Feng, 2003: 24).

15 Patent Law, Article 58; Trademark Law, Article 59; Copyright Law, Article 47.

16 This coded case is based on an interview with a senior MNC executive.

17 Provisions of the Supreme People's Court on issues relating to adjudication of patent disputes, Article 21.

18 TRIPS agreement, Article 13; Regulation on Protection of Computer Software, Article 17.

19 This case is based on interviews with a senior executive and an independent IP attorney in China. For confidentiality reasons the company's name has been coded.

20 Sildenafil citrate is the main ingredient in the popular drug Viagra for the treatment of erectile dysfunction.

21 The technical information disclosed in the description should be able to achieve the effect described in the claim (Patent Law, Article 26).

22 Also called a confidential disclosure agreement or a confidentiality agreement.

23 If no special equipment is involved, it is sometimes very difficult for an employer to prove the use of corporate materials and technological means.

24 Restricted technology: approval of the Ministry of Commerce and royalty payments are required for export (Regulation on Restriction of Technology Import and Export, Article 33).

25 Paris Convention for the Protection of Industrial Property, Article 4.

8

Managing cooperation with local universities

The idea of seeking cooperation for R&D endeavours is not new. Companies from the triad regions of the world have long been engaged in R&D cooperation with universities, research institutes and other entities (Hagedoorn and Narula, 1995). In China, the increase in R&D internationalisation that has resulted in emerging R&D cooperation has been apparent for at least a decade. According to recent research on the growth of international R&D alliances in China (Li and Zhong, 2003), different forms of R&D cooperation[1] between MNCs and local Chinese partners, such as companies, universities, research institutes and government agencies, steadily rose in the period between 1995 and 2000 (Figure 8.1). Due to the Chinese government's loosening of regulatory policies, the joint venture as a form of R&D cooperation declined in favour of cooperative R&D alliances between foreign R&D investors and Chinese domestic parties.

Previous research has indicated that local universities and research institutions are preferred above other local cooperation partners by international R&D organisations. Hence, the focal point of this part of the research will be R&D cooperation between foreign R&D organisations and local universities and research institutes.[2] Largely due to IP

Figure 8.1 **International R&D alliances in China (1995–2000)**

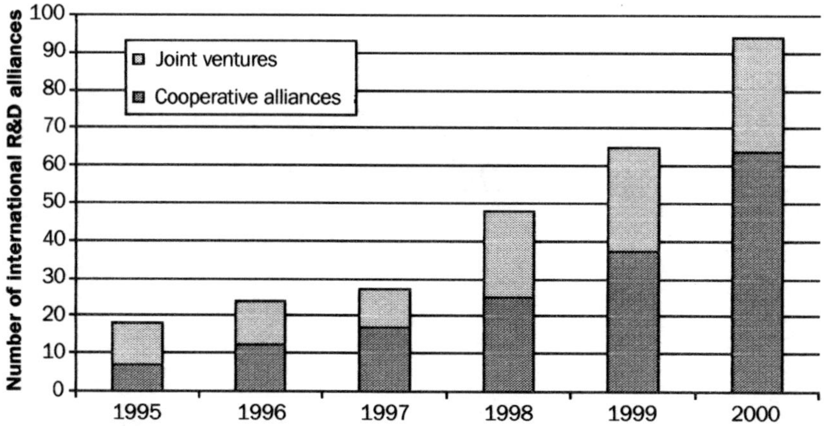

Source: Li and Zhong (2003)

concerns and a lack of R&D competency, local companies are less favoured as cooperation partners in the R&D area.[3] This type of cooperation is therefore excluded from the research focus.

Although R&D cooperation could turn out to be very fruitful for all the involved parties, this area still lacks experience, particularly in the China context. For historical reasons, many universities and research institutes do not have much experience in cooperating with industry partners, especially not with partners from abroad. On the other hand, MNC R&D organisations are quite familiar with industry-university cooperation, but their experience has been gained within a context totally different from that in China. In 2003 a study by the Chinese Academy of Social Sciences found that around 60 per cent of the surveyed foreign-invested companies had no experience of working with the local Chinese governments and 77 per cent had not worked with Chinese research institutes (ZDNet Asia,

2003). It is therefore a reasonable presumption that both parties harbour uncertainties regarding the other.

The research in this chapter aims to contribute to the understanding of managing R&D cooperation between foreign R&D organisations and local Chinese universities and research institutes. First, the chapter reviews the existing literature on industry-university R&D cooperation. Based on the preceding field research and the identified literature gap, research questions will be formulated. Second, in-depth research requires an understanding of the context of R&D cooperation in China. Hence, China's science and technology (S&T) system and policy will be introduced. Third, based on the field research results and selected case studies, the research questions will be answered and the ensuing managerial implications developed.

The context of R&D cooperation

To manage R&D cooperation with universities and research institutes successfully, one first needs to have a comprehensive understanding of the Chinese S&T system and the different government policies and measures that promote joint R&D between industries and research organisations. In this section, the most important development indicators of China's national R&D activities and driving S&T policies are outlined for an understanding of some of the trends and developments in recent years. This is followed by a discussion of the Chinese S&T system's organisational structure, and identification of the key stakeholders and their functions in relation to R&D cooperation.

Indicators of national R&D activities in China

As many indicators suggest, China's R&D has made steady progress over the last 25 years. The country's spending on R&D has shown a continuous increase in the last decade. Between 1998 and 2003 its gross expenditure on R&D nearly tripled from RMB551.1 million to RMB1.5 billion, which was 1.31 per cent of GDP in 2003 (Figure 8.2) (China Science and Technology Statistics, 2004). The objective of the Tenth Five-Year Plan (for the period 2001–2005) was to increase overall spending on R&D to 1.5 per cent of GDP (*People's Daily*, 2001). It is likely that China will achieve this goal. The rapid development in R&D expenditure allowed this developing country to approach the level of the world's most developed and technologically advanced economies in the triad regions, which usually spend 2–3 per cent of their GDP on R&D.

According to China Science and Technology Statistics (2004), three major sectors do R&D work in China: enterprises (62.4 per cent), research institutes (25.9 per cent) and universities (10.5 per cent). Similar to other countries, most R&D is done by industry. Most of China's R&D

Figure 8.2 **National R&D expenditure (1991–2003)**

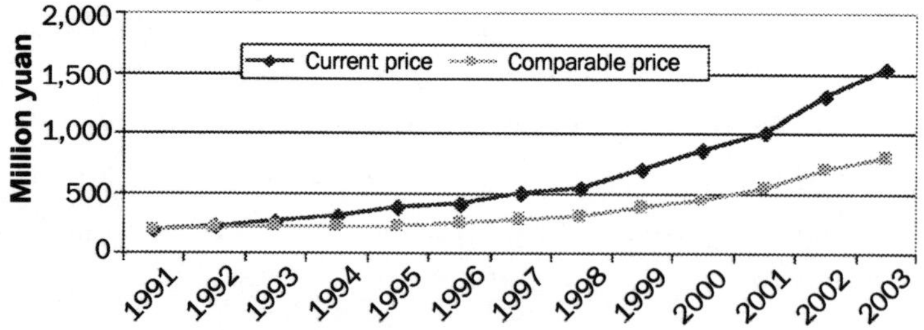

Source: China Science and Technology Statistics (2004)

spending is allocated to technology development (74.1 per cent); applied research constitutes about 20 per cent; and basic research accounts for only 5.7 per cent.

A number of other R&D indicators show that China is making steady progress in its high-tech pursuits (Table 8.1). According to the Science Citation Index (SCI), China ranked sixth in 2003 in terms of the number of scientific papers contributed (with a total of 49,788 papers, which represents 4.48 per cent of the world total) (China Science and Technology Statistics, 2004). In terms of the SCI, the Index to Scientific & Technical Proceedings (ISTP) and the Engineering Index (EI) as a whole, China ranked fifth in the number of scientific papers. Similarly, the number of scientific publications in domestic journals has also risen steadily from 133,341 in 1998 to 274,438 in 2003.

Due to the Chinese government's emphasis on the funding of technology development and applied research activities, the numbers of patent applications and grants continue to

Table 8.1 World ranking of S&T paper output (2003)

Country	SCI, ISTP and EI			SCI		
	Papers	%	Rank	Papers	%	Rank
China	93,352	5.09	5	49,788	4.48	6
USA	564,090	30.74	1	359,610	32.36	1
Japan	154,241	8.41	2	92,448	8.32	3
UK	139,911	7.62	3	96,280	8.66	2
Germany	131,688	7.18	4	85,591	7.70	4
France	89,854	4.90	6	59,762	5.38	5
Italy	71,858	3.92	7	45,882	4.13	7
Canada	68,296	3.72	8	44,829	4.03	8
Russia	44,437	2.42	10	26,968	2.43	11
India	34,136	1.86	14	23,136	2.08	13

Source: China Science and Technology Statistics (2004)

grow. The total number of patent applications quadrupled and patents granted in the last decade grew more than threefold (Hu and Jefferson, 2005). The majority of patents are applied for by and granted to enterprises. Institutions for higher education and state research institutes are improving their productivity in terms of patent output and, together, have as many patents granted as industry (China Science and Technology Statistics, 2004). Yet the same statistics also indicate that the degree of true innovation remains low, as most of the patents granted are not for inventions, but for designs or utility models. Although there are three times more domestic patent applications than foreign patent applications in China, the number of foreign invention patent applications is almost equal to that of domestic applications. This clearly indicates that domestic R&D capability in China is more development- than research-oriented, as shown previously (Hu and Jefferson, 2005).

After 20 years of implementing market-driven reforms to modernise its S&T system, China's economy is becoming more technologically advanced, as indicated in Figure 8.3 by the growing volume of Chinese high-tech exports.

According to World Bank (2002) data, however, much of this is contributed by foreign-invested enterprises, which

Figure 8.3 **National exports of high-tech products (1993–2003)**

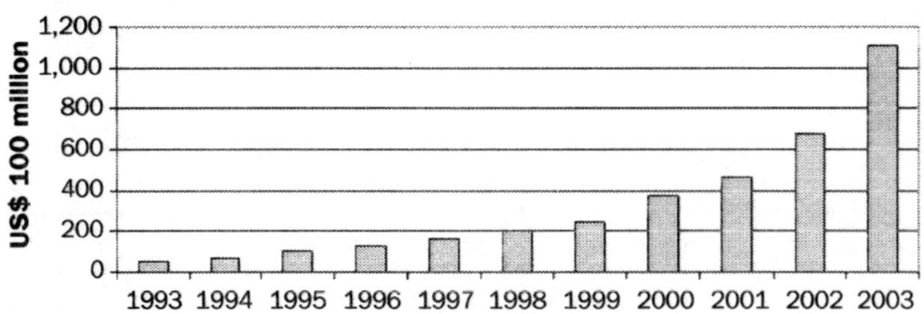

Source: China Science and Technology Statistics (2004)

produce nearly 50 per cent of China's total exports. Nevertheless, the country's ultimate ambition is to accelerate its high-tech development and be among the world's most competitive high-tech economies. The above statistics reveal that the country is unquestionably making strides to reach its goals. The institutional S&T framework that enables China's continuing advancement is described in the next section.

Institutional framework of the Chinese S&T system

To understand how China aims to develop its S&T system, one must understand its organisational structure. After experiencing decades of reform and reorganisation with the gradual slackening of the planning mechanism, the Chinese S&T system's current institutional framework is maturing and becoming more comprehensive. It has different ministries, foundations and research institutions as major stakeholders (see Figure 8.4). The State Council's Steering Committee of S&T and Education and the Leading Group on Science, Technology and Education are the highest-ranking S&T policy coordination bodies. Headed by the State Council premier, the steering committee carries out coordination of the government policy and national strategy for S&T and education. It also oversees S&T policy in the ministry and at the local level.

The Ministry of Science and Technology (MOST) is responsible for coordinating and organising all official Chinese S&T activities. MOST implements several national programmes to fund S&T development, with three focusing specifically on R&D: the 863 National High-Technology R&D Programme, the 973 National Basic Research

Figure 8.4 **Organisational structure of the Chinese S&T system**

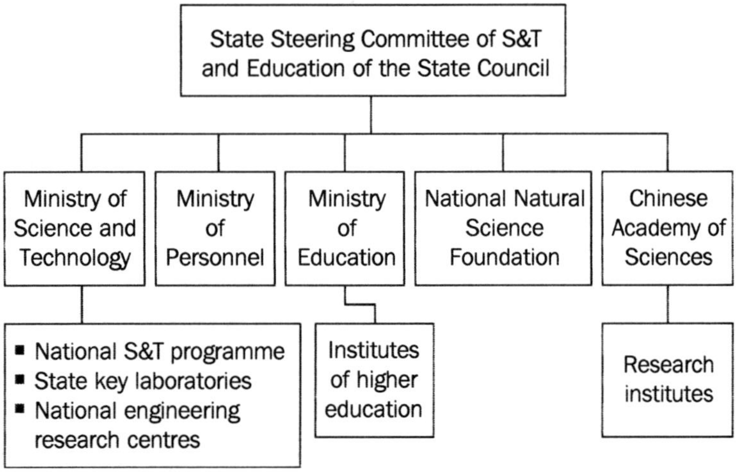

Source: Based on Huang et al. (2004)

Priorities Programme and the Torch Programme (Ministry of Science and Technology, 2005a).

In terms of involvement in the national S&T system, the Ministry of Personnel, besides being responsible for the management of China's post-doctoral system, is, together with the Ministry of Education, also responsible for developing measures to persuade expatriate overseas students and scholars to return to China (Huang et al., 2004).

The Ministry of Education (MOE) sets the guiding principles and policies for China's educational undertakings. Moreover, it cooperates with other Chinese organisations to develop a series of funding programmes, such as the Overseas Returnee Scientific Research Fund, to recruit Chinese research talent around the world to work in China (Ministry of Education, 2004). This is accomplished by awarding leading researchers who excel in certain key areas.

The National Natural Science Foundation of China (NSFC) promotes and finances S&T research and is modelled on the US National Science Foundation. It funds basic and applied research in natural sciences. The grant awardees are mostly Chinese universities and CAS research institutes (Huang et al., 2004).[4]

The Chinese Academy of Sciences (CAS) is China's foremost natural science and technology research organisation and operates 89 independent research institutes that function as a research extension of the Chinese government (Chinese Academy of Sciences, 2004). CAS members and institutes serve as consultants to the government, providing S&T policy advice. It is thus obvious that the CAS is the major beneficiary of government funding for S&T.[5]

China's S&T policy

The institutional framework depicted above maintains the country's S&T policy, which will be illustrated in more depth in this section. An analytical scheme proposed by Huang et al. (2004) was adopted and modified (see Figure 8.5) to provide an appropriate analysis of the S&T policy.

Science and technology policy

Chinese S&T policy's current objective is to accelerate high-tech development through 'informatisation', a technology development programme that encompasses all sectors of China's economy. In China's Tenth Five-Year Plan (2001–2005), an investment of more than 1 trillion RMB was proposed in the IT industry, with the goal of doubling its size to comprise a full 7 per cent of GDP (Walsh, 2004).

Figure 8.5 An analytical framework of China's S&T policy

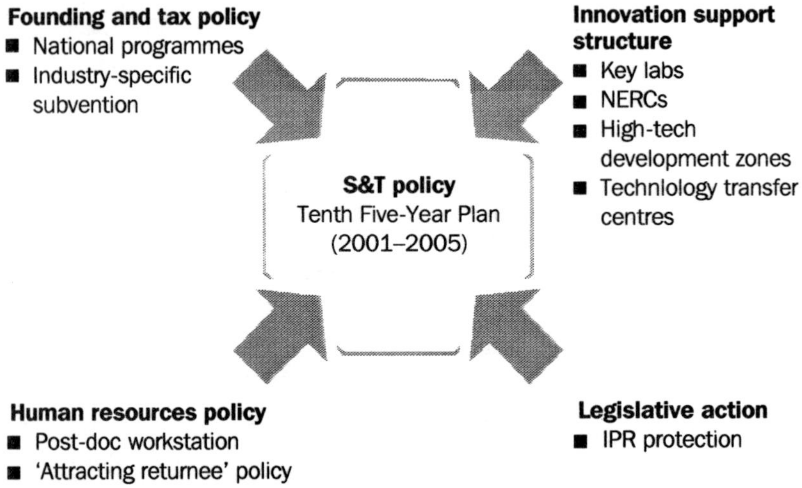

Source: Based on Huang et al. (2004)

The S&T policy also seeks to prioritise spending on 'pillar industries' and key strategic technologies (including IT and electronics), and to reform China's SOE R&D system to enable further increases in economic development.[6] Local governments play a decisive role in executing the planned S&T policy. For example, following the release of 'Several Policies to Encourage the Development of Software and Integrated Circuit Industries' by the central government in 2000, the municipal government of Shanghai took a series of steps to develop the semiconductor industry in the area (UNCTAD, 2005).

Funding and tax policy

In the 1980s China started a series of national programmes to fund R&D activities (see Table 8.2). In 2004 the Key Technologies R&D Programme,[7] the 863 Programme[8] and

Table 8.2 **China's current S&T programmes**

Programme	Initiated	Objective
Key Technology R&D Programme	1983	Concentrate resources on key and common technologies urgently needed by developments for upgrading industries and social sustainability
State Key Laboratories Programme	1984	Support selected laboratories at public or private facilities
Spark Programme	1986	Support technology transfer to rural areas to promote rural development
National Natural Science Foundation	1986	Support basic research by funding projects directly
863 High-Tech R&D Programme	1986	Enhance China's international competitiveness and improve overall capability regarding high-tech R&D
Torch Programme	1988	Support high-tech industry sector development by setting up science parks and incubators, funding projects, human resource training etc.
973 Key Basic Science R&D Programme	1997	Support basic science research
Innovation Fund for Small Technology-based Firms	1999	Support establishment of new technology-based firms

Source: Based on Huang et al. (2004: 373)

the 973 Programme[9] emerged as the biggest funding programmes led by MOST, attracting 72 per cent of the funds managed by MOST for R&D in the country (Huang et al., 2004).

Besides the major national programmes initiated by the government, the Chinese authorities have continuously implemented preferential tax policies targeted at foreign investors, especially in the high-tech manufacturing and service sectors. For example, for a long time semiconductor companies in China enjoyed a value-added tax rebate favouring chips made and/or designed in the country through a regulation (Document 18) of China's Ministry of Information Industry (MII). Despite the fact that Document 18 had to be cancelled on 1 April 2005 because it violated WTO principles, the MII released new regulations in 2005 to support China's semiconductor industry by providing a subsidy fund of RMB100–200 million (*EETimes*, 2005).

Innovation support structure

To strengthen its technological infrastructure, in 1992 China borrowed US$14 million from the World Bank to establish 55 state key laboratories for basic research and 14 national engineering research centres (NERCs) (US Department of Commerce, 2000). State key laboratories are centres of excellence in research, and have greatly promoted China's basic research activities and overall S&T development (*Asia Pacific Biotech News*, 1999). NERCs promote applied research and engineering in priority industries such as electronics and telecommunications; they are designed to link up the research conducted by mostly CAS-affiliated institutes to manufacturing processes and production.

By 2002 over 400 business incubators and 53 high-technology development zones had been established through governmental support at the national level, mainly thanks to the Torch Programme. These zones created around 3.49 million new jobs and spent RMB31.47 billion on R&D, which was 24.4 per cent of the gross expenditure on R&D in China and 40 per cent of the business expenditure on R&D (Ministry of Science and Technology, 2003). Furthermore, according to the OECD 'Science, technology and industry outlook' (OECD, 2004), one major measure in promoting the linkage between industries and scientific research in China was the establishment of national technology transfer centres. They will enhance the alliance between industries, universities and research institutes, and advance the combination of university and industry S&T resources, as well as the transfer of advanced proven technologies to industries.

Human resources policy

China has continually increased its investment in education. During the Ninth Five-Year Plan (1996–2000) the total national expenditure on education increased by an average of 15.4 per cent annually. By the end of the Tenth Five-Year Plan the total expenditure on education could reach 5 per cent of GDP, compared to 2.59 per cent in 1998 (China Education Yearbook, 2002). Besides further investment, the MOE has focused on the steady diffusion of higher education. In the past few years China has increased the percentage of high-school graduates having access to higher education institutions from 4 per cent to a gross enrolment rate of 19 per cent. In 2005 the total on-campus student population exceeded 20 million at more than 2,000 HEIs (Ministry of Education, 2005).

At the postgraduate level, China has also caught up with the world's leading countries. In 2002 China ranked third behind the USA and Germany in terms of annual PhD awards (*People's Daily*, 2002c). Within the next few years, China could surpass the USA and become the biggest 'producer' of PhDs. During the past two decades, 1,363 mobile post-doctoral stations have been established at Chinese universities and research institutions and 1,018 in industries (Ministry of Science and Technology, 2005b).

The government has also recognised that an important source of high-quality personnel is Chinese students abroad.

As Figure 8.6 reveals, the number of students returning from abroad has increased steadily since the early 1990s. The thriving Chinese economy and slowly liberalising political environment have transformed China's brain drain into brain gain since the economy's opening, and the Chinese government has invested tremendous efforts in attracting overseas Chinese academics and business people back to their country of origin (see Zweig, 2005). New government organisations and numerous programmes have been established to advance this goal, which has many implications in terms of R&D cooperation with universities and research institutes.

Legislative action

Besides economic reform, China has introduced comprehensive legislative reform and released a series of S&T-, education- and IPR-related laws. The IPR-related laws and legislation are critical for S&T and innovation. Beijing has clearly recognised that strengthening IPR protection for Chinese innovators and a sharper focus on IP management benefit the country itself while further

Figure 8.6 **Number of returned students (1990–2003)**

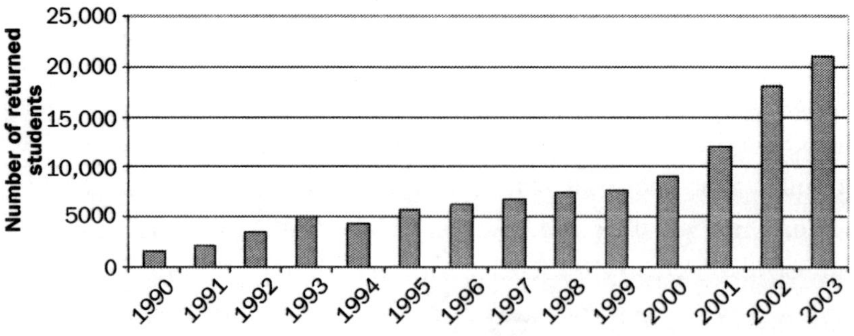

Source: China Science and Technology Statistics (2004)

developing its S&T. As discussed in Chapter 7, IP is a rather recent concept in China, and its IP system has only been created during the last two decades. Because of its relatively short history, the enforcement of IPR protection has been difficult due to a number of structural, political and cultural factors. Consequently, it is critical for MNCs wishing to engage in R&D cooperation to understand China's IP laws, their implementation, the Chinese court system and how to deal with infringements.

Although the preceding sections have shown that China is progressing by leaps and bounds, the country still has a long way to go to catch up with the leading industrialised nations. Huang et al. (2004) compared Chinese S&T policy with that of the EU, using the European Trend Chart on Innovation,[10] and also contrasted China's S&T performance with that of OECD countries by using the OECD Science, Technology and Industry Scoreboard.[11] They maintain that, due to a lack of policy coordination, the potential for enhancing China's innovation performance remains untapped in areas such as education, R&D enterprises and cooperation between research institutions, universities and companies.

Case examples of MNC cooperation with local universities

Having considered some indicators of R&D activities in China, as well as its S&T system, organisational structure and policies, this section now introduces three selected case examples of cooperation between MNCs and local universities and/or research institutions, based on field research. From different perspectives, these cases can be considered best practice.

Lucent Technologies – Bell Labs Research China

Lucent Technologies is a leading solution provider for communications networks. Bell Labs – the innovation arm of Lucent Technologies – is the largest R&D organisation in the world. It is focused on the needs of service providers in the telecommunications industry and conducts fundamental research in fields of new communications technologies and networking (Bell Labs, 2004). As the first Bell laboratory established outside the USA, Bell Labs Research China (BLRC) was launched in 2000 and assigned to support Lucent's business units and customers in the China and Asia-Pacific region.

BLRC set up its first branches in Beijing and Shanghai, and joint labs with China's prestigious institutions of higher education – Peking University and Shanghai Jiaotong University – in 1997. Currently, BLRC is maintaining six joint research labs with leading Chinese universities and research institutes[12] that engage in research on optical networks, next-generation internet IPv6, wireless communications, software systems and analysis and design of networks. BLRC also has project-based direct cooperation with professors on specific research topics and runs a post-doctoral programme at the research centres. The setting of the joint research lab with Shanghai Jiaotong University is exemplary: BLRC provides all the resources, i.e. equipment, funding (except salaries), training and research topics, while the Chinese partner provides the researchers and space. Generally, these labs are led by professors in specific areas and include several key BLRC members. In Lucent's experience, approaching a university's top decision-makers directly and establishing strong personal ties to leading professors are both key to launching a joint project successfully. In terms of IP regulations, detailed contract agreements are always negotiated in advance. In most cases, the IPRs resulting from the cooperation belong to BLRC.

For the time being, BLRC does not actively participate in national programmes. However, it does have a funding agreement with the NSFC to fund jointly some of the NSFC's major research and young scholar foundation projects. Since the telecommunication industry is strongly dependent on industry standards and policies, it is crucial to have intensive dialogue with the governmental authorities. The Chinese government has established different kinds of standards workgroups, and BLRC is active in the wireless standards workgroup – this participation is essential to the company's product development.

Unilever

In 2003 Unilever, a leading supplier of consumer goods, opened the Unilever Research China centre in Shanghai, focusing on developing detergents, hair, bathing, skin and oral products, beverages such as tea and condiments like soy sauce products for the Chinese market. The R&D centre also has core R&D capabilities in areas such as synthetic materials, micro-structured particles (nanotechnology) and TCM. An important building-block in the formation of the R&D centre was the company's previous cooperation with Chinese research institutions. To 'test the water', a joint lab for synthetic research was founded at the Shanghai Institute of Organic Chemistry, a CAS institution, in 1996. This was an R&D joint venture focusing on synthetic materials, a field in which Chinese scientists are very strong. The CAS provided the R&D staff and premises; and Unilever provided the funding and research topic and also led the project management. Moreover, Unilever also cooperates with professors on specific research projects and, from time to time, outsources its non-core capability in cooperation with universities and/or research institutes.

Unilever applies systematic mapping to identify cooperation partners, whether they are universities or individual professors, according to reputation and academic publications. Besides approaching the ideal cooperation partner directly, Unilever also sends out calls for research proposals in order to filter the most appropriate cooperation partners. Again, detailed contractual agreements on IP-related issues are fundamental for smooth R&D cooperation. If a project's outcome is identified as having high commercial potential, Unilever will take over the supplementary R&D once the outcome is worthy of the company's capacity and capability. All academic articles are checked before publication to ensure that important knowledge and IP created through cooperation will not be leaked.

Unilever helps university research projects that have commercial potential to apply for national programme funding, helps universities to participate in discussions and provides advice regarding research from a product development perspective. The benefit for the company is that it obtains early access to research results with commercial value. For example, Unilever encourages universities to apply for the Shanghai Unilever Research and Development Fund, established by Unilever with the Shanghai Science and Technology Commission, that supports projects in the life sciences. It funds universities and research institutes in the Shanghai area in fundamental and applied research of mutual interest to the Shanghai municipality and Unilever; it also provides scholarships for selected PhD candidates and research fellows.[13]

Nokia

Nokia, a world leader in mobile communications, started to manufacture its products in China in the mid-1990s. In 1998 Nokia established its Nokia China R&D centre in Beijing. Almost simultaneously, the company started to work with local universities. Since then the Finnish company has built seven R&D units in different Chinese cities, such as Beijing, Shanghai, Hangzhou and Chengdu. It has a wide range of cooperation ventures with local universities (Vaario, 2005):

- sponsoring university research projects for emerging technology;
- scholarships, innovation funds, equipment, conference participation;
- transfer of knowledge and technical training;
- subcontracting projects;
- joint national projects (863 and NSFC) as a non-funded partner, EU projects;
- academic information exchange;
- internship positions and joint post-doctoral programmes.

In general, Nokia does not have joint research labs with local universities. Joint research projects remain the main form of cooperation at the technical level. Through these different cooperation activities, Nokia helps local universities to develop through exposure to cutting-edge technology and international trends; in return, the company's high visibility helps it to receive high-quality contributions from talented local scientists. To date more than 60 projects have funded more than 100 researchers and students, leading to more than 100 publications (Vaario,

2005).[14] Due to their innovative and win-win character, the following two cooperation endeavours need to be highlighted in more detail.

In response to strong demand, Nokia offers training and seminars on mobile application development for university teaching staff and students. Currently, six universities in Beijing follow the programme and more universities from other locations will join in future. This kind of training promotes and transfers specific, practically relevant knowledge and competence that Chinese universities lack. Simultaneously, Nokia benefits from the increasing number of skilled application developers. Three universities have already developed several mobile applications.

Nokia also provides a knowledge innovation fund at the Beijing University of Posts and Telecommunications for graduate students' research and thesis work, promoting and enhancing innovation consciousness and performance. It is not only very helpful for Nokia to identify young talent, but it also generates a positive PR effect at a personal level. Nokia's experience is summarised as 'little money for Nokia, but a big impact on an individual' (Vaario, 2005: 11).

Discussion of research findings

The subsequent four sections discuss the findings regarding the respective research questions, i.e. the form of R&D cooperation, partner selection and partnership establishment practice, IP management and national programme participation, and the practical managerial implications.

Major motives

The motivations for R&D cooperation with universities and research institutes have a significant impact on the forms of cooperation, which in turn affect the selection of a partner and the cooperation orientation's objectives. Table 8.3 lists the major motives for industrial firms to cooperate with universities, as identified in the existing literature.

According to the field research findings, it is obvious that the motives for firms to cooperate with universities and research institutes in China correspond to some of the motives listed in the literature from the international

Table 8.3 **Major motives for industrial firms to cooperate with universities**

Motives	Related research
Obtain knowledge regarding basic research and emerging technologies	Elsevier, 1989; Tseng, 1990; Geisler et al., 1991; Roessner et al., 1998
Directly obtain technical support	Geisler et al., 1991; Roessner et al., 1998
Increase interaction with other organisations and/or technical staff	Smilor, 1987; Tseng, 1990; Roessner et al., 1998
Reduce costs and risks	Liu, 1983; Chen, 1985; Elsevier, 1989; Tseng, 1990
Shorten learning curves	Liu, 1983; Smilor, 1987
Use expensive facilities at universities	Liu, 1983; Elsevier, 1989
Gain important knowledge about competitors	Elsevier, 1989
Stimulate R&D creativity	Liu, 1983; Spark, 1985; Elsevier, 1989
Recruit bright students as future employees	Liu, 1983; Spark, 1985; Elsevier, 1989; Tseng, 1990; Roessner et al., 1998

Source: Wu (2001: 4)

perspective. Three types of motives have, however, been identified as most significant in the Chinese context: to establish early contacts with universities regarding talent; obtain specialised competence, knowledge of basic research and emerging technologies; and intensify interaction with the local scientific community and obtain access to information through networks.

Talent recruitment is one of the most important motives for MNCs to cooperate with local universities. Since foreign R&D investment has rapidly increased over the last years, the demand for high-profile graduates is also soaring. The fight for top talent at prominent Chinese universities such as Tsinghua, Peking and Fudan is getting more intense. R&D cooperation is therefore often used as a platform to enable regular interaction and dialogue with the next generation of top graduates. Cooperation with local universities and research institutes has thus become a critical HR recruiting instrument for MNC R&D organisations in China.

MNCs are keen to acquire specialised competency that complements their own R&D efforts. Some MNCs even fully outsource non-core capabilities to universities or research institutes, and directly purchase their specialised competencies or IP, such as bacterial stems in the biotech industry. In addition, outsourcing certain R&D work to a university is considered an effective way to bridge internal capacity bottlenecks. Furthermore, the interviewed R&D managers commonly mentioned that working closely with universities enables them to obtain cutting-edge knowledge of basic research that can facilitate the identification of emerging technologies and support the defining of a future R&D direction. This is more pertinent in disciplines such as biotechnology and chemistry in which the local scientific community has strong R&D capabilities.

Networking is critical in the Chinese environment. The dynamism of R&D in China is not only determined by technological advancement, but also by governmental policy and regulation changes, which are very fluid. R&D cooperation with local universities and research organisations offers an ideal platform for tapping into the local scientific community, which is familiar with the policy environment and has better access to first-hand regulatory information.

At the same time, however, there is a certain discrepancy due to the specificity of the Chinese environment. Reducing costs and risks, gaining important knowledge about competitors and stimulating R&D creativity were merely mentioned as important motives in the interviews. This may reflect that, beyond a few top-flight institutions, the majority of Chinese universities are still not sufficiently mature to produce world-class R&D innovation and creativity. As a motivation, gaining knowledge about competitors means a competitor could have an equal chance of learning more about you. The risk of losing information to a competitor is even higher within the Chinese context, as personal networks could sometimes outweigh written non-disclosure agreements. As discussed in Chapter 6, Chinese universities' strength does not lie in being creative.

In conclusion, while MNCs' cooperation with universities and research institutes could lead to knowledge regarding basic research and emerging technology, technical support, human resources and specialised competencies becoming available, they should also expect to contribute to equipment and other necessities, such as knowledge, to facilitate advanced R&D.

Forms of cooperation

In this research, the analysis of forms of cooperation between MNCs and local universities or research institutions is carried out from two different perspectives: outcome- and relationship-focused cooperation.

Outcome-focused R&D cooperation

For outcome-focused cooperation, MNCs primarily expect to achieve clear technological outputs, e.g. product development, product innovations and even patents. Joint research labs, joint research projects and post-doctoral programmes are examples of this kind of cooperation. Outcome-focused cooperation has a positive side-effect: long-term relationship-building. This will be tackled in more depth in the next section.

Joint research projects are less equity-based and thus reflect less organisational interdependence and more flexibility. A joint research project usually has a clearly identified topic and limited timeframe. Both parties contribute and share according to a previously negotiated project contract. Due to the flexibility and low equity involvement, MNCs often first choose to cooperate with local universities and research institutions in this form of cooperation in order for them to get to know each other better. Based on this experience, a decision is made regarding more in-depth cooperation. R&D subcontracts can be considered an extreme form of a joint research project. Sometimes MNCs fully outsource their non-core competencies to local partners and use universities as service providers.

The field research indicates that a joint research lab is the most common equity-based cooperation type between

MNCs and local universities. Equity-based cooperation involves equity investments, and thus reflects a tighter organisational interdependency between the partners. As a rule, big MNCs choose to establish joint research labs with prominent universities in order to secure long-term cooperation; this form of cooperation often does not have a legal entity in China. In addition, equity-based cooperation is usually established in industry sectors in which high investments in terms of equipment are required for R&D activities. As mentioned before, lack of resources is a common phenomenon among local universities and research institutions; thus establishing a joint research lab usually means that the university provides the research staff and (often) the location, and the foreign partner provides the equipment and funding. Such a constellation allows MNCs a stronger influence in determining the research focus. The local universities benefit in that they are then able to offer high-quality training opportunities of higher practical relevance to their students and researchers through their close interaction with the industry. There are huge differences between joint research labs in terms of scale and scope. As one interviewed R&D manager pointed out, a few joint research labs have a higher PR effect than their real output – a phenomenon which is typical within the Chinese context.

R&D joint ventures that take the form of a legal person are relatively rare. This kind of R&D cooperation is more frequently found in formerly or still regulated sectors such as the automobile industry (*The Economist*, 1999; Walsh, 2003). Inflexibility, lack of control, the risk of IP leakage and a lack of equal mutual benefits are among the major concerns restraining MNCs from R&D joint ventures. In other words, the transaction cost of this type of cooperation is much too high.

Another outcome-oriented cooperation form is post-doctoral workstations. Through these workstations, a company seeks to gain benefits from high-quality publications, or even patents and inventions. When firms decide to participate in China's post-doctoral programme and establish a workstation at their R&D organisation, they can either gain a mandate to run a post-doctoral workstation through direct application (which needs to be approved by the Ministry of Personnel) or they can choose to cooperate with a university that has such a mandate. For example, Nokia set up a programme offering long-term (five to ten years) research funding for post-doctorates in 2004. The programme serves as a platform for the exploration of emerging technologies, with a focus on specific topics such as mobile multimedia applications and future terminal and networking technologies.

Field research revealed several success factors in respect of these kinds of cooperation. First, the project target must be concrete, with clear deliverables that require a realistic estimation of mutual capabilities. Second, as the project leader, an MNC must provide sufficient coordination and control the resources available. The success of long-term R&D endeavours very much depends on the highest management level's continuous support. Third, having the appropriate (preferably the best) university and professor as partners is crucial for productive cooperation. Regarding long-term cooperation, many of the interviewed managers and external experts from academia emphasise the importance of cooperating with the right professors. This issue is discussed in more depth below.

Relationship-focused cooperation

The second category of cooperation identified by the field research is characterised by its strong relationship-building orientation and usually does not involve equity critical for R&D activities. Sponsorship in the form of scholarships and innovation funds, transfer of knowledge through training, lectures and advisory activities at universities (see below) and internship programmes are relationship-focused cooperation activities. Scholars used to interpret non-equity-based cooperation by using the resource-based view of the firm according to which firms try to access other parties' various resources (see Hagedoorn, 2001; Palmberg and Martikainen, 2003). Without a doubt, and particularly within the Chinese context, a relationship can be regarded as a type of critical resource required to succeed. MNCs can build, strengthen and maintain their links to local universities and their association with top students or future employees through investment in the form of knowledge transfer and funds. Besides the PR effect, talent recruitment and relationship-building with authorities, this kind of investment yields other benefits. Via knowledge transfer in the form of special training, MNCs better prepare their potential future employees for their practical requirements. Furthermore, in for example IT industries, free training also helps companies to make best use of certain software. Different cooperation forms thus benefit both parties in different ways.

Partner selection

As pointed out in the previous section, working with the right cooperation partners is key for MNCs to succeed in industry-university cooperation. Logically, the country's best scientists and faculties are at the best universities, such as Tsinghua, Fudan, Zhejiang etc. Hence, these top-flight universities will be the most appropriate candidates for cooperation. However, not only is the capacity at leading universities in China limited, but the number of professors producing work at the international level is also very limited. Due to the increasing demand for R&D cooperation, foreign and domestic companies are competing with each other for the limited resources at these famous universities. Following the free-market rule, top universities in China are raising the threshold for cooperation, e.g. the minimum investment for a joint research lab. MNCs with big names and large investments have a greater advantage than companies that are less well known and not as financially strong. Yet top faculty and professors are not only available at well-known universities in the country: potential at 'second-tier' universities, as they are called, is mainly untapped. Especially for smaller MNCs with fewer resources available, it is of strategic importance for their R&D to tap this potential. The major challenge for MNC R&D managers is to find the right cooperation partner. Mapping and evaluating a potential partner's S&T capability is the first step. Here the correct way of identifying and evaluating the right school and/or professor for cooperation is critical. The author provides a model that could assist MNCs to identify the right school for cooperation (see Table 8.4).[15] With minor modifications, this tool could be used to evaluate a single professor as well as the school as a whole. Most of the information required for the evaluation is publicly available or accessible with a little research effort. Most of the suggested evaluation categories are quantifiable. The potential cooperation partner's ability and qualifications can therefore be objectively estimated.

As discussed previously, China's S&T policy seeks to invest intensively in certain high-tech areas and build centres of excellence. Alliances with such competence and knowledge hubs will yield advantages for MNCs in terms of

Table 8.4 **Evaluation tool for potential cooperation partners**

Category	Criteria	Candidate A	Candidate B	Candidate C
Quality of S&T capability	Number of academic publications (national/international) Number of patent applications Number of granted patents			
S&T capacity	Number of industry-specific professors/ researchers Number of PhD and masters students in relevant field			
Vertical linkage	Number of industry cooperations with manufacturers			
Horizontal linkage	Number of alliances with other R&D organisations			
Government relationships	Number of professors with governmental obligations National programme participation			
Relation to market	Public reputation of university Number of commercialised patents or innovations			
Equipment	Relevant technology facilities			

the availability of qualified researchers and better-equipped laboratories. Moreover, a local university's participation in national programmes is also a quality indicator in the cooperation partner's evaluation. For example, in 2003 the

MOE initiated the China National IC Design Talent Incubation Project, China's first national IC design training programme.[16] Nine universities were designated for participation: many are not first-tier universities, but were chosen by the government to demonstrate the second-tier universities' underlying potential.

Many of the above-discussed cooperation forms are relatively costly. Smaller foreign companies can leverage R&D cooperation more effectively and cost-efficiently through long-term cooperation with a few selected top professors. As discussed in Chapter 6, a relationship with top professors is also a very reliable channel for recruiting good R&D professionals.

Intellectual property management

The interview findings reveal that a detailed contract agreement in terms of fully owning, sharing or licensing IP is always negotiated at the start of cooperation. Most of the interviewed MNCs insist on having the IPRs of all the project outcomes. The final settlement of the IP issue depends on negotiations and both parties' relative positions. Needless to say, the more input a party invests, the stronger its bargaining position. According to several of the interviewees, the leading universities have already gained some experience in dealing with the IP issue through their previous cooperation with foreign partners. On the other hand, the less-experienced universities are often over-aware of the IP issue and afraid that MNCs could benefit more from cooperation and take advantage of them. Furthermore, as Chinese universities' management is becoming more performance-oriented, i.e. a faculty's success is measured by indicators such as publications and patents, university officials may interfere in the IP issue in order to emphasise their university's interest. MNCs could deal with such situations by using a well-established international cooperation contract as a template or starting point for further IP negotiations. A practical interim solution mentioned during the field research is to list the specific professor as the inventor in respect of a patent application, but with renouncement of the commercial claim.

As discussed in the last chapter, some firms have also developed their own IP management strategy. For example, joint research projects that have been identified as having a high commercial value potential will be returned to the company if the company has the capacity and capability to work on its own. Most interviewed high-tech firms report that they have not encountered any IP infringement through R&D cooperation with their university partners. However, as IP disputes are always a very sensitive topic, the interviewees may not have shared all their experiences with the researcher, and especially not the negative ones relating to existing relationships.

National programme participation

A question many MNC R&D managers often ask is to what extent MNCs can participate in national S&T programmes such as 863 or 973. Although the Chinese government officially actively encourages universities and industries to apply for national programme funding together, in practice it is almost impossible for foreign companies to be funded or become directly involved, due to a lack of governmental support. An external expert from academia revealed that China faces a high demand for funding but has a limited R&D fund capacity, so funding is primarily given to Chinese domestic industries and universities. MNCs are expected to make a contribution to the national programmes rather than take advantage of them.

It is apparent that a few MNCs have succeeded in becoming a kind of non-funded partner. Others participate more indirectly by supporting university research projects with commercial value in order to apply for national programme funding by helping universities to participate in discussions and provide research advice from an MNC's specific experience, e.g. from a product development perspective. The benefit of this kind of non-funded or indirect participation is that it could ensure early access to research results or priority rights in respect of commercialisation. In addition, participation in national programmes is to a certain degree still dependent on *guan xi*. Having said this, cultivating key relationships might help foreign firms tap into the national programmes' funding.

In several of the interviewees' experience, MNCs are far more likely to become involved in the National New Products Programme. This is an annual programme that supports R&D efforts in new high-tech products, particularly those with a high degree of innovativeness and locally developed IP. The programme preferentially supports the commercialisation of the 863 Programme's research results, domestic manufacturers' products with high export potential and products compatible with or promoting international standards. Participants receive a certificate, a financial subsidy and preferential tax reductions or exemptions, and the new product is published on the programme's website (National New Products Programme, 2005). The major value of participating in the National New Products Programme does not lie in the financial aid, which is relatively low, but the reputation and official recognition of the MNCs' commitment to and integration into the Chinese economy.

Conclusion

The successful management of R&D cooperation between a company and a university could yield mutual benefits. With the increasing liberalisation of the economy and the ongoing reform of the Chinese S&T system, this topic is becoming increasingly relevant in China. Working closely with local universities and research institutions can be particularly beneficial for MNCs in many ways. Nonetheless, contextual and cultural unfamiliarity may steer scarce resources and managerial attention away from a firm's core strategy (Peng and Heath, 1996). Thus international R&D organisations in China need to overcome uncertainties by making themselves familiar with the context first, i.e. the Chinese S&T system and policy. Furthermore, the research has pointed out the following key areas to which international R&D units in China need to pay more attention, especially during the initial stage of cooperation.

Which form of cooperation MNCs should adopt depends on the foreign R&D organisation's underlying motives. Outcome-focused R&D collaboration could help MNCs to leverage their competency and become more productive and innovative. Relationship-oriented cooperation is very helpful for HR recruitment and networking, which are key determinants for successful R&D activities in China. For smaller companies, it will be more effective and cost-efficient to work closely with individual professors.

Finding the right partner is the cornerstone of a promising collaboration endeavour. The research has suggested a partner-mapping tool that could help MNCs to identify and evaluate potential university cooperation partners. Furthermore, given the inadequate institutional infrastructure in China's transitional economy, cultivating close connections with different authorities can be an effective way of gaining resources and influence for firms. Working with the right partner is again key.

The interviewees did not consider the IP infringements resulting from collaborations with local universities as serious as in other areas. Besides the IP-related issues discussed in the last chapter, the research findings urge MNCs to pay more attention to IP ownership regulations before starting cooperation. IP ownership should always be thoroughly discussed and contractually agreed in order to avoid unpleasant conflicts that could harm the good relationship and waste resources. Usually, each party's input determines their part of the IP output.

Although participation in national programmes could yield multiple benefits, it is almost impossible for foreign parties to be directly involved. Yet there are several ways for international R&D organisations to benefit indirectly from these programmes, e.g. through access to the priority rights of new technology's commercialisation, and by taking advantage of the funded labs through close collaboration. At the same time, the government also expects inputs, particularly in the form of transfer of knowledge from the MNCs.

Notes

1 In research by Li and Zhong (2003), cooperation alliances are identified as research projects, joint product developments, R&D joint ventures and R&D consortia (excluding vertical supplier partnerships and one-way technology licensing agreements, technology support and training agreements).

2 In China the majority of research institutes are state-owned.

3 A study by the Chinese Academy of Social Sciences in 2003 found that 79 per cent of surveyed foreign-invested companies saw no reason to work with domestic enterprises (ZDNet Asia, 2003).

4 In general, the NSFC offers grants in five different programmes: the Key Programme, the Major Programme, the National Science Fund for Distinguished Young Scholars, the Science Fund for Innovative Research Groups and the Special Funds (National Natural Science Foundation, 2005).

5 In 2003 22 per cent of the total funding of China's NSFC went to the CAS. It successfully applied for more than one-third of the 52 projects of Programme 973 for the years 2002 and 2003. In 2001 and 2002 it received 14.1 per cent of Programme 863's total funding. In 2003, of the 333 key national basic research projects, 92 were undertaken by CAS institutions (Chinese Academy of Sciences, 2003, 2004).

6 Twelve key technologies were identified as the S&T focus during the Tenth Five-Year Plan (Walsh, 2004).

7 This programme is based on the targets of the Tenth Five-Year Plan. It aims to catch up with advanced economies in agriculture and key industry sectors, and emphasises the development of environmental technology and strengthening the private sector in order for it to be a major technological innovator (Key Technologies R&D Programme, 2004).

8 This programme supports the development of applied research in civilian and military fields such as biotechnology, information, automation, energy, advanced materials, marine and space and laser (see 863 Programme, 2004).

9 Contrary to the 863 Programme, this programme only provides funds for basic research in civilian areas of science such as energy resources, information resources, environment, and population and health. It aims to ensure a scientific foundation for innovative research, promote scientists and establish high-level research units (973 Programme, 2004).

10 The European Trend Chart on Innovation is the European Commission's benchmarking instrument in the area of innovation policy.

11 The Science, Technology and Industry Scoreboard brings together the latest internationally comparable data on trends in the knowledge-based economy. The analysis covers four key areas: creation and diffusion of knowledge, information society, economic globalisation and productivity and economic structure (OECD, 2004).

12 Such as Tsinghua University, Peking University, Fudan University and the CAS.
13 See also www.rdfund.sh.cn.
14 Jari Vaario was director of technology at Nokia China from 2001 to 2005.
15 One interviewed MNC has developed a kind of internal intelligence or screening system for identifying the most appropriate universities or professors with which to cooperate. However, access to this system was not granted to the researcher. The model developed here is inspired by interview findings.
16 The programme was developed by the MOE and Cadence Design Systems, the world's largest electronic design automation (EDA) technologies and engineering services company. The government's target is to train 300 students per year per university at master and doctorate levels. Cadence will provide wide-ranging support for the initiative, from IC design and EDA courses to jointly developed design projects (Cadence, 2003).

9

Conclusion and outlook

This chapter concludes the discussions and findings of both research phases and highlights the relevance of the work. In addition, it underlines the limitations of the study as well as the potential for future research in this area.

The emergence of foreign R&D activity in China is a recent phenomenon and has seen rapid growth since China's WTO accession. As a non-traditional locus for international R&D activities, the dynamic Chinese environment, which is characterised by high uncertainties, remains a huge challenge for many MNCs in managing their R&D activity. The two-phase, in-depth field research endeavoured to gain more insights into the phenomenon. The first research phase was specifically dedicated to two questions: why China is tremendously attractive for MNC R&D activities, and what kind of challenges and uncertainties they face within this new and dynamic context.

The field research uncovered a series of determinants of MNC motivations for establishing R&D in China. The author examined the identified drivers by using an analytical cluster made up of three categories of motivation: performance-oriented, business-ecology-oriented and input-oriented. Achieving customer- and market-oriented development for a market with tremendous potential with the help of qualified personnel who are mostly available are the two most important incentives for foreign R&D to enter

China. Home-base-exploiting R&D activities still dominate home-base-augmenting motivations, but since the country's scientific maturity is increasing steadily it will not be surprising if more HBA R&D activities appear in China in the near future. The research also pointed out the important role of business-ecological determinants, such as governmental interference and relationship-building aspects, which have been largely ignored in the prevailing HBA and HBE theory.

The more advantages, the higher the risks: this rule of thumb also applies to localising R&D activities in China, and the considerable challenges could outweigh the potential benefits. Three areas of challenges surfaced as the most critical: managing R&D professionals, intellectual property and industry-university cooperation. This finding led to the focus of the second research phase: exploring the core challenges in more depth. In order to close the distance between the researcher and the proceedings under investigation, a participation observation method was applied. Regular on-site observations and interviews with relevant employees of a newly established foreign R&D centre allowed valuable first-hand data to be collected.

Chapter 6 started the discussion of the management of R&D staff with an examination of the strengths, deficits and particularities of young professionals and graduates in China; this lays the essential groundwork for further analysis. For institutional and cultural reasons, the very motivated young Chinese R&D professionals possess both high potential and particular weaknesses. The research indicated that context-adjusted management practices and proper training and coaching opportunities will allow MNCs to release Chinese R&D staff's huge underlying capability. The field research further discovered that recruiting talent based on a firm's *guan xi* network in China could be an advantageous approach. In order to retain talent in the long run, which is the most critical issue facing foreign R&D centres, this cannot be achieved by relying solely on monetary incentives and attention must be paid to a mixture of factors. As the leadership largely determines a culturally diversified organisation's success, appointing the right managers will be key. In the light of increasing staff localisation, global Chinese who embody the characteristics associated with a 'glocalpreneur' could be the most suitable candidates for taking over the leadership of foreign R&D centres in China.

Chapter 7 tackled the management of IP in China, a well-known struggle for all MNCs, by first building a solid understanding of the context, i.e. the IPR protection environment and the IP protection legal system in China. The research then underlined that MNCs need to be particularly aware of several external sources of uncertainties over which they have only limited, if any, control, such as the weak awareness of IP, loopholes in the IP administrative and legislative systems, local protectionism and corruption. Important managerial implications were developed for domains such as IP ownership, IP leakage, IP ownership disputes, IP-related corporate coordination and IP transfer, during which uncertainties can be avoided by appropriate managerial steps. The awareness of problematic issues, the robustness of the local IP ownership and the building of strategic relationships should be acknowledged as key for MNCs' management of IP-related uncertainties in China.

As pointed out in Chapter 8, cooperation between universities and industry has only a relatively short history in China. The existing contextual and cultural unfamiliarity makes strategy-driven preparations critical for leveraging mutual benefits. The choice of cooperation form should be determined by the underlying motives. The research identified two types of industry-university cooperation: outcome-focused collaborations that help MNCs to accelerate R&D productivity, and relationship-oriented collaborations that mainly aim to ensure a supply of high-quality HR and important networks. Due to high demand for the limited cooperation capacity at well-known universities, MNCs with limited financial means and less-known brand names could have more disadvantages than 'big-name' companies. MNCs can leverage certain competitive advantages by identifying promising potential and establishing cooperation with excellent second-tier universities. For this purpose, a directly applicable tool for evaluating local universities was presented.

With this book, the author hopes to add to the understanding of the drivers of and barriers to foreign R&D activities in China. The findings of research phase II could specifically be highly relevant for MNCs in managing their R&D engagements in China. The identified managerial implications are based on the valuable expertise of the interviewed R&D managers and executives, as well as specialised IP attorneys. Furthermore, the opinions of government officials and independent field experts from academia and consultancies contribute to an objective, balanced view of the focal research fields. The findings could enable foreign as well as Chinese R&D managers to build up a holistic view to deal with particular uncertainties and challenges in their daily business. The awareness generated may help them to take a further step towards being a 'glocalpreneur'.

There are still many challenges ahead for MNC R&D activities in China. The following strategic issues should have first priority on their agenda. First, it is of strategic importance to secure knowledge generated by local R&D

centres in the long run. Secondly, in order to achieve a deeper integration into MNCs' global R&D networks, it will be critical for them to enhance local units' R&D productivity and quality towards the international level. Simultaneously, MNCs need to keep updating their internal strategy of R&D internationalisation, while taking their medium- and long-term purpose in China as well as the development progress of their local R&D units into consideration.

The research also has its limitations. Firstly, due to the dynamic environment in China, the validity of the findings could be limited in terms of longevity. The conclusions and research findings should not be taken for granted without paying attention to recent developments. Secondly, because only one manager or executive of each company was interviewed in most cases, the reliability of the statements and anecdotal information could be biased by subjectivity. Furthermore, as some interviewees had only recently been assigned to their positions, their experiences could be preliminary. Thirdly, major parts of the research (except Chapter 8) are practice-oriented and intended to generate constructive managerial implications. Future research could increase the theoretical integration of the findings. Lastly, given the exploratory nature of the research, only qualitative methods were used, which limits the generalisability of the findings. At the time the field research started, there were less than 200 known foreign R&D organisations in China and many of these had just launched their activities, which was one reason why quantitative research methods were not considered. Taking a realistic average response rate into account, it was not expected that a sufficient number of replies would be received to allow statistical analysis methods to be applied. However, during the field research the number of foreign R&D organisations in China rose to over 800, which opens new opportunities for research by means of quantitative methods. This research can thus be considered a preparatory work for future endeavours of this kind.

Due to the challenging nature of the context, different complementary disciplines such as HRM and cooperation management have become critical success factors in managing R&D in China. Future research domains of high practical and theoretical relevance therefore lie in the intersecting areas of different management disciplines.

Appendix: IP-related laws and regulations in China

Agreement on Trade-related Aspects of Intellectual Property Rights (TRIPS)

Paris Convention for Protection of Industrial Property

Several Provisions of the Supreme People's Court on Issues Relating to Application of Law to Adjudication of Cases of Patent Disputes

PRC Arbitration Law

PRC Copyright Law

PRC Implementing Regulation of Copyright Law

PRC Implementing Regulation of Patent Law

PRC Law Against Unfair Competition

PRC Patent Law

PRC Regulation on Computer Software Protection

PRC Regulation on Protection of IC Layout Designs

PRC Regulation on Restriction of Technology Import and Export

PRC Trademark Law

References

Alford, P.W. (1995) *To Steal a Book Is an Elegant Offence – Intellectual Property Law in Chinese Civilization.* Palo Alto: Stanford University Press.

Ambrecht, F.M.R. (2002) 'WTO entry, government's welcome could spur foreign R&D in China', *Research and Technology Management*, 45(5): 2–5.

Asakawa, K. (1996) 'External-internal linkages and overseas autonomy-control tension: the management dilemma of the Japanese R&D in Europe', *IEEE Transactions on Engineering Management*, 43(1): 24–32.

Asakawa, K. (2001) 'Organisational tension in international R&D management: the case of Japanese firms', *Research Policy*, 30: 735–57.

Asia Pacific Biotech News (1999) 'National key laboratories – a success in China'; available at: *www.asiabiotech.com/03/english/preserved-docs/0321/0520_0521.pdf.*

Asia Pulse (2002) 'Taiwan's R&D should differentiate from that of China', *Asia Pulse*, 5 December.

AT Kearney (2004) 'FDI confidence index 2004'; available at: *www.atkearney.com/shared_res/pdf/FDICIOct_2004_S.pdf.*

Bai, J.B. and Cheng, H. (2004) 'Chinese patents'; available at: *www.ipfrontline.com/downloads/Chinese_patents.pdf.*

Beckmann, C. and Fischer, J. (1994) 'Einflussfaktoren auf die Internationalisierung von Forschung und Entwicklung in der Chemischen und Pharmazeutischen Industrie', *Zeitschrift für Betriebswirtschaftliche Forschung*, 46(7/8): 630–57.

Beise, M. and Belitz, H. (1996) 'Internationalizierung von F&E multinationaler Unternehmen in Deutschland', in Gassmann, O. and von Zedtwitz, M. (eds) *Internationales Innovations Management*. Munich: Vahlen, pp. 215–30.

Bell Labs (2004) 'Bell Labs Research China'; available at: *http://blrc.edu.cn/blrcweb/about.htm#blrc*.

Boutellier, R., Gassmann, O. and von Zedtwitz, M. (2000) *Managing Global Innovation: Uncovering the Secrets of Future Competitiveness*, 2nd edn. Berlin: Springer.

Boutellier, R., Kloth, B. and Bodmer, C. (1996) 'Neue Organisationsformen globaler Forschung und Entwicklung', *Zeitschrift Führung und Organisation*, 65(5): 282–7.

Brockhoff, K. and Guan, J. (1996) 'Innovation via new ventures as a conversion strategy for the Chinese defence industry', *R&D Management*, 26(1): 49–56.

Cadence (2003) 'Cadence and China's Ministry of Education sign landmark MOU to build IC design talent base in China'; available at: *www.cadence.com/company/newsroom/press_releases/pr.aspx?xml=112603_beijing*.

Cantwell, J. (1992) 'The internationalization of technological activity and its implications for competitiveness', in Granstrand, O., Hakanson, L. and Sjölander, S. (eds) *Technology Management and International Business*. London: John Wiley & Sons, pp. 75–95.

Chen, H. (2005) 'Managing intellectual property in China', *Kind & Wood IP Bulletin*, 3: 1–9.

Chen, I.L. (1985) 'Harnessing university research for competitiveness, industry support', *IEEE Spectrum*, 27(10): 73–6.

CH-ina (2005a) 'Human resource management', in Musy, S., Horvath, S. and Rechsteiner, A. (eds) *Behind the China Kaleidoscope – A Guide to China Entry and Operations*. Shanghai: CH-ina (Shanghai), pp. 468–93.

CH-ina (2005b) 'Intellectual property protection and R&D', in Musy, S., Horvath, S. and Rechsteiner, A. (eds) *Behind the China Kaleidoscope – A Guide to China Entry and Operations*. Shanghai: CH-ina (Shanghai), pp. 402–9.

China Daily (2004a) 'President: China targets USD 4 trillion GDP by 2020'; available at: *www.chinadaily.com.cn/english/doc/2005-05/16/content_442681.htm*.

China Daily (2004b) 'Huawei, Cisco settle IPR dispute'; available at: *http://www2.chinadaily.com.cn/english/doc/2004-07/30/content_353070.htm*.

China Daily (2005) 'Chinese firms file lawsuit on DVD patent'; available at: *http://www2.chinadaily.com.cn/english/doc/2005-01/20/content_410667.htm*.

China Daily (2007) 'Trade surplus hits record after 74 percent growth'; available at: *www.chinadaily.com.cn/china/2007-01/26/content_793150.htm*.

China Economic Review (2005) 'Investment activity (May 15 – Jun 14, 2005)', *Foreign Investment Report*, 3(7): 3–25.

China Education Yearbook (2002) *China Education Yearbook 2002*. Beijing: People's Education Press.

China Information Industry Net (2005) 'Chinese mobile phone subscribers reaches 377 million'; available at: *www.cnii.com.cn/20050801/ca319726.htm*.

China IPR News (2005) 'IPR – the first four years of the Tenth Five Year Plan' (in Chinese), *China IPR News*, 31 January.

China Legal Information Centre (2005) 'Analysis of a dispute on service invention' (in Chinese); available at: *www.law-star.com/pshowtxt?keywords=&dbn=lwk&fn=01-2001-6-76.txt&upd=1.*

China Science and Technology Statistics (2003) 'Characteristics and trends of MNCs' R&D organisations in China' (in Chinese); available at: *www.sts.org.cn/tjbg/other/documents/2003/0301.htm.*

China Science and Technology Statistics (2004) 'China science & technology statistics 2004'; available at: *www.sts.org.cn/sjkl/kjtjdt/data2004/cstsm04.htm.*

China Statistics Bureau (2001) *Statistics of China's Science and Technology in 2000. China Social Statistics*. Beijing: China Statistics Press.

Chinese Academy of Sciences (2003) 'Scientific and research activities, Chinese Academy of Sciences 2002 annual report' (in Chinese); available at: *www.cas.ac.cn/html/Books/O6121/b1/2002/5.pdf.*

Chinese Academy of Sciences (2004) 'Scientific and research activities, Chinese Academy of Sciences 2003 annual report' (in Chinese); available at: *www.cas.ac.cn/html/Books/O6121/b1/2003/Index.htm.*

Clark, D. (2000) 'IP rights protection will improve in China – eventually', *China Business Review*, 27(3): 22–8.

Dalton, D.H. and Serapio, M.G. (1995) *Globalizing Industrial Research and Development*. Washington, DC: US Department of Commerce.

Davey, G. and Fei, Y. (2005) 'Meeting the needs of Chinese students on overseas private higher education programs: Chinese students' self-evaluation of their study skills',

International Journal of Private Higher Education, 1: 25–31.

De Boer, S.J., Gan, W. and Shan, G. (1998) 'Critical issues facing R&D managers in China', *R&D Management*, 28(3): 187–97.

Deng, L. (2004) 'How to protect software-related inventions in China', *Managing Intellectual Property*, 142: 20–4.

Dunning, J.H. and Narula, R. (1995) 'The R&D activities of foreign firms in the United States', *International Studies of Management & Organisation*, 26: 461–91.

EETimes (2005) 'China set to launch chip subsidy fund, says report'; available at: *www.eetimes.com/showArticle.jhtml?articleID=166404120*.

Elsevier (1989) 'Motivations and obstacles to R&D management', *Technovation*, 9: 161–8.

Feng, P. (2003) *Intellectual Property in China*. Hong Kong: Sweet & Maxwell Asia.

Fischer, W. (1983) 'The structure and organization of Chinese industrial R&D activities', *R&D Management*, 13(2): 63–81.

Fischer, W. and Farr, C.M. (1985) 'Dimensions of innovative climate in Chinese R&D units', *R&D Management*, 15(3): 183–91.

Fischer, W. and von Zedtwitz, M. (2004) 'Chinese R&D: naissance, renaissance or mirage?', *R&D Management*, 34(4): 349–65.

Florida, R. (1997) 'The globalization of R&D: results of a survey of foreign-affiliated R&D laboratories in the USA', *Research Policy*, 26(1): 85–103.

Gassmann, O. (1997) *Internationales F&E-Management: Potentiale und Gestaltungskonzepte transnationaler F&E-Projekte*. Munich and Vienna: Oldenbourg.

Gassmann, O. and Han, Z. (2004) 'Motivations and barriers of foreign R&D activities in China', *R&D Management*, 34(4): 423–37.

Gassmann, O. and von Zedtwitz, M. (1998) 'Organization of industrial R&D on a global scale', *R&D Management*, 28(3): 147–61.

Gassmann, O. and von Zedtwitz, M. (1999) 'New concepts and trends in international R&D organization', *Research Policy*, 28: 231–50.

Geisler, E., Furino, A. and Kiresuk, J.T. (1991) 'Toward a conceptual model of cooperative research: patterns of development and success in university-industry alliances', *IEEE Transactions on Engineering Management*, 38(2): 136–45.

Gelb, C. (2000) 'Software research in and for China: Microsoft's basic research facility takes global R&D a step further', *China Business Review*, 27(4): 40–3.

Gerpott, T. (1990) 'Globales F&E-Management: Bausteine eines Gesamtkonzeptes zur Gestaltung einer weltweiten F&E-Organisation', *Die Unternehmung*, 4(44): 226–46.

Gerybadze, A. and Reger, G. (1999) 'Globalization of R&D: recent changes in the management of innovation in transnational corporations', *Research Policy*, 28: 251–74.

Granstrand, O. (1999) 'Internationalization of corporate R&D: a study of Japanese and Swedish corporations', *Research Policy*, 28(2/3): 275–302.

Granstrand, O., Håkanson, L. and Sjölander, S. (eds) (1992) *Technology Management and International Business: Internationalization of R&D and Technology*. Chichester: John Wiley & Sons.

Granstrand, O., Håkanson, L. and Sjölander, S. (1993) 'Internationalization of R&D – a survey of some recent research', *Research Policy*, 22: 413–30.

Greatwall Strategy Consultants (2002) *R&D Embracing China* (in Chinese). Nannin: Guangxi Renmin Publishing.

Hagedoorn, J. (2001) *Inter-firm R&D Partnerships – An Overview of Major Trends and Patterns Since 1960.* Eindhoven: Eindhoven Centre for Innovation Studies.

Hagedoorn, J. and Narula, R. (1995) 'Choosing organizational modes of strategic technology partnering: international and sectoral differences', *Journal of International Business Studies*, 2: 265–84.

Håkanson, L. (1992) 'Locational determinants of foreign R&D in Swedish multinationals', in Granstrand, O., Håkanson, L. and Sjölander, S. (eds) *Technology Management and International Business: Internationalization of R&D and Technology.* Chichester: John Wiley & Sons, pp. 97–115.

Håkanson, L. and Nobel, R. (1993a) 'Determinants of foreign R&D in Swedish multinationals', *Research Policy*, 22: 397–411.

Håkanson, L. and Nobel, R. (1993b) 'Foreign research and development in Swedish multinationals', *Research Policy*, 22: 373–96.

Håkanson, L. and Zander, U. (1988) 'International management of R&D: the Swedish experience', *R&D Management*, 18(3): 217–26.

Han, Z., von Zedtwitz, M. and Gassmann, O. (2005) 'International R&D in China – opportunities and risks', in Musy, S., Horvath, S. and Rechsteiner, A. (eds) *Behind the China Kaleidoscope – A Guide to China Entry and Operations*. Shanghai: CH-ina (Shanghai), pp. 418–27.

Heakal, R. (2003) 'What is an emerging market economy?'; available at: *www.investopedia.com/articles/03/073003.asp*.

Helble, Y. (2004) *Corporate Research and Development in a Late Industrializing Context – The Case of Singapore*. Bamberg: Difo.

Hilb, M. (2002) *Transnationales Management der Human Ressourcen*. Neuwied/Kriftel: Luchterhand Verlag.

Hilb, M. (2004) *Integriertes Personal-Management: Ziele – Strategien – Instrumente* (Chinese edition). Beijing: China Labour and Social Security Publishing House.

Hofstede, G. (1991) *Cultures and Organisations: Software of the Mind*. London: McGraw-Hill.

Hofstede, G. (1994) *Cultures and Organisations – Intercultural Cooperation and its Importance for Survival*. London: HarperCollins.

Hofstede, G. (2003) *Culture's Consequences, Comparing Values, Behaviours, Institutions, and Organizations Across Nations*, 2nd edn. Newbury Park, CA: Sage Publications.

Howells, J. (1990a) 'The location and organisation of research and development: new horizons', *Research Policy*, 19: 133–46.

Howells, J. (1990b) 'The internationalization of R&D and the development of global research networks', *Regional Studies*, 24: 495–512.

HR Weekly (2004) 'China lacks patent examiners – 300 planned recruitments in 2005' (in Chinese), *HR Weekly*, 17 September.

Hu, A. and Jefferson, G.H. (2005) 'A great wall of patents: what is behind China's recent patent explosion?', discussion paper; available at: *http://people.brandeis.edu/~jefferso/Great%20Wall,%20submitted%20manuscript,%20Jan.%202006.pdf.*

Huang, C., Amorim, C., Spinoglio, M., Gouveia, B. and Medina, A. (2004) 'Organisation, programme and structure: an analysis of the Chinese innovation policy framework', *R&D Management*, 34(4): 367–87.

Huawei (2005) 'Robust R&D'; available at: *www.huawei.com/about/AboutListDoc.do?webColType=abouthuawei&root_id=ROOT|C233|C425|C442&docId=ROOT|C233|C425|C442|C956.*

IIPA (2003) 'International Intellectual Property Alliance 2002 special 301 report on global copyright protection and enforcement – People's Republic of China'; available at: *www.iipa.com/rbc/2002/2002SPEC301PRC.pdf.*

James, G. (2004) 'Fighting for your rights – how bad is IP theft in China and what can you do about it?', *Electronic Business*, August: 58–61.

Jiao, S., Ji, G. and Jing, Q. (1986) 'Comparative study of behavioural qualities of only children and sibling children', *Child Development*, 57: 357–61.

Jin, S. and Porter, A. (1988) 'Technological innovation and development: prospects for China', *IEEE Transactions on Engineering Management*, 35(4): 258–64.

Johanson, J. and Vahlne, J.E. (1977) 'The internationalization process of the firm: a model of knowledge development and increasing foreign commitments', *Journal of International Business Studies*, 8: 23–32.

Jolly, D.R. (2004) 'Bartering technology for local resources in exogamic Sino-foreign joint ventures', *R&D Management*, 34(4): 389–406.

Jossi, F. (2000) 'Challenge in China', *Human Resource Executive*, 2 October: 106.

Judicial Protection of IPR in China (2005) 'Recent developments of patent enforcement in China (3) – a comprehensive legal framework at work'; available at: *www.chinaiprlaw.com/english/news/news33.htm.*

Kenney, M. and Florida, R. (1994) 'The organization and geography of Japanese R&D: results from a survey of Japanese electronics and biotechnology firms', *Research Policy*, 23: 305–23.

Key Technologies R&D Programme (2004) 'Key Technologies R&D Programme in the Tenth Five-Year Plan' (in Chinese); available at: *www.gongguan.most.gov.cn/intro105.htm*.

Kogut, B. and Chang, S. (1991) 'Technological capabilities and foreign direct investment in the United States', *Review of Economics and Statistics*, 73(3): 401–13.

Kuemmerle, W. (1999) 'The drivers of foreign direct investment into research and development: an empirical investigation', *Journal of International Business Studies*, 30(1): 1–24.

Le Bas, C. and Sierra, C. (2002) 'Location versus home country advantages in R&D activities: some further results on multinationals' localization strategies', *Research Policy*, 31: 589–609.

Legislation Working Group of National People's Congress (2003) *Compilations of Laws and Regulations of People's Republic of China* (in Chinese). Beijing: Law Press.

Levinthal, D.A. and Cohen, W.H. (1990) 'Absorptive capacity: a new perspective on learning and innovation', *Administrative Science Quarterly*, 35(1): 128–52.

Li, H. and Atuahene-Gima, K. (2001) 'Product innovation strategy and the performance of new technology ventures in China', *Academy of Management Journal*, 44(6): 1123–34.

Li, J. and Zhong, J. (2003) 'Explaining the growth of international R&D alliances in China', *Managerial and Decision Economics*, 24 (special issue): 101–5.

Li, Z.S. (2005) 'IP protection in China', *Kind & Wood IP Bulletin*, 3: 1–4.

Liu, B.C. (2003) 'Discovering Chinese cultural roots', in Li, Y. and Reuvid, J. (eds) *Doing Business with China*. London: GMB Publishing, pp. 175–81.

Liu, J.L. (1983) *How Corporations Cooperate with Academic Organizations for R&D*. Taipei: Tsong-Te Industrial Research Foundation.

Liu, X. and White, S. (2001) 'Comparing innovation systems: a framework and application to China's transitional context', *Research Policy*, 30: 1091–1114.

Mansfield, E., Teece, D. and Romeo, A. (1979) 'Overseas research and development by U.S. based firms', *Economica*, 46: 187–96.

McComb, R. (1999) 'China's human resource odyssey', *China Business Review*, 26(5): 30–3.

Medcof, J.W. (2001) 'Resource-based strategy and managerial power in networks of internationally dispersed technology units', *Strategic Management Journal*, 22(11): 999–1012.

Melvin, S. (2000) 'Human resources take center stage', *China Business Review*, 27(6): 38–49.

Ministry of Education (2003a) *Report on the Development of Countrywide Educational Undertaking in 2002*. Beijing: Ministry of Education.

Ministry of Education (2003b) *Short Introduction to Studying and Working Abroad. Ministry Report*. Beijing: Ministry of Education.

Ministry of Education (2004) 'Work related to students and scholars studying abroad'; available at: *www.moe.edu.cn/english/international_2.htm*.

Ministry of Education (2005) 'The 9th 5-Year Plan for China's educational development and the development outline by 2010'; available at: *www.moe.edu.cn/english/planning_n.htm*.

Ministry of Science and Technology (2003) 'National high-technology zones'; available at: *www.most.gov.cn/l_1_tt.htm#d08*.

Ministry of Science and Technology (2005a) 'About MOST – the Ministry of Science and Technology'; available at: *www.most.gov.cn/English/about/about_main.htm*.

Ministry of Science and Technology (2005b) 'Newsletter of 2005.08.04'; available at: *www.most.gov.cn/eng/newsletters/2005/t20050804_23687.htm*.

Moga, T. and Jonathan, R. (2002) 'The TRIPS agreement and China', *China Business Review*, 29(6): 12–19.

Mutinelli, M. and Piscitello, L. (1998) 'The entry mode choice of MNEs: an evolutionary approach', *Research Policy*, 27: 491–506.

National Bureau of Statistics of China (2005) *China Statistical Yearbook 2004 – National Bureau of Statistics Plan Report*. Beijing: National Bureau of Statistics of China.

National Natural Science Foundation (2005) 'Guide to programme'; available at: *www.nsfc.gov.cn/e_nsfc/2004/02gp/index.htm*.

National New Products Programme (2005) 'The major supports of National New Products Programme' (in Chinese); available at: *www.chinanp.gov.cn/jhgs7.htm*.

National Science Board (2004) *Science & Engineering Indicators – 2004*. Arlington, VA: National Science Foundation.

Neumann, P.A. (2005) 'An introduction to intellectual property protection in China', in Musy, S., Horvath, S. and Rechsteiner, A. (eds) *Behind the China Kaleidoscope – A Guide to China Entry and Operations*. Shanghai: CH-ina (Shanghai).

Niosi, J. (1997) 'The globalization of Canadian R&D', *Management International Review*, 37(4): 387–404.

Nokia (2005) 'Nokia opens mobile infrastructure R&D center in Sichuan in China'; available at: *www.nokiausa.com/about/newsroom/article/1,1046,3412,00.html.*

Odagiri, H. and Yasuda, H. (1996) 'The determinants of overseas R&D by Japanese firms: an empirical study at the industry and company levels', *Research Policy*, 25: 1059–79.

OECD (1998) *Internationalization of Industrial R&D: Pattern and Trends*. Paris: OECD.

OECD (2002a) *Proposed Standard Practice for Surveys on Research and Experimental Development (by Frascati Manual)*. Paris: OECD.

OECD (2002b) *OECD Science, Technology and Industry Outlook 2002*. Paris: OECD.

OECD (2003a) OECD *Science, Technology and Industry Scoreboard 2003*. Paris: OECD.

OECD (2003b) *OECD Investment Policy Reviews – China: Progress and Reform Challenges*. Paris: OECD.

OECD (2004) 'Science, technology and industry outlook 2004 – Country response for policy questionnaire: China'; available at: *www.oecd.org/dataoecd/31/27/34241720.pdf.*

Ong, L. (2004) 'China's banking system a ticking time bomb'; available at: *www.atimes.com/atimes/China/FA13Ad02.html.*

Palmberg, C. and Martikainen, O. (2003) *The Economics of Strategic R&D Alliances – A Review with Focus on the ICT Sector*. Helsinki: Research Institute of the Finnish Economy.

Papanastassiou, M. and Pearce, R.D. (1994) 'The internationalization of research and development by Japanese enterprises', *R&D Management*, 24(2): 155–65.

Patel, P. and Vega, M. (1999) 'Patterns of internationalization of corporate technology: location vs home country advantages', *Research Policy*, 28: 145–55.

Pearce, P. and Pooni, G.S. (1996) 'The globalization of R&D in pharmaceuticals, chemicals and biotechnology: some new evidence', discussion paper, International Investment and Business Studies, Series B, No. 213, pp. 1–33.

Pearce, R.D. (1989) *The Internationalization of Research and Development by Multinational Enterprises*. New York: St Martin's Press.

Pearce, R.D. and Papanastassiou, M. (1996) 'Overseas R&D and the strategic evolution of MNEs: evidence from laboratories in the U.K.', University of Reading Working Paper Series B, Vol. IX, No. 225.

Peng, M.W. and Heath, P.S. (1996) 'The growth of the firm in planned economies in transition: institutions, organisations, and strategic choice', *Academy of Management Review*, 21: 492–528.

People's Daily (2001) '"10th FYP": China's overall national strength expected to rank world's 5th'; available at: *http://english.people.com.cn/english/200104/03/eng20010403_66711.html.*

People's Daily (2002a) 'China tipped to become world hi-tech center'; available at: *http://english.people.com.cn/200206/03/eng20020603_97074.shtml.*

People's Daily (2002b) 'China's foreign trade sets new high despite adverse circumstances'; available at: *http://english.people.com.cn/200210/28/eng20021028_105789.shtml*.

People's Daily (2002c) 'China may have more PhDs than the US in 2010'; available at: *http://english.people.com.cn/200210/21/eng20021021_105393.shtml.*

People's Daily (2003a) 'Chinese universities to enroll 3.35 million students in 2003'; available at: *http://english.people.com.cn/200301/16/eng20030116_110216.shtml.*

People's Daily (2003b) '80 percent of 2002 college graduates find jobs'; available at: *http:// english.people.com.cn/200301/23/eng20030123_110644.shtml.*

People's Daily (2003c) 'Three-year-old conflict revived again' (in Chinese); available at: *www.people.com.cn/GB/it/1066/2262395.html.*

People's Daily (2004a) 'Foreign R&D centers in China on the rise'; available at: *http://english.people.com.cn/200408/20/eng20040820_153974.html.*

People's Daily (2004b) 'Trade secrets stealing – the Huawei case' (in Chinese); available at: *www.people.com.cn/GB/shehui/1061/3041021.html.*

People's Daily (2004c) 'Authorities: Viagra patent found invalid'; available at: *http://english.people.com.cn/200407/09/eng20040709_149013.html.*

People's Daily (2005a) 'Market share of Chinese home-branded mobile phones continued to drop: survey'; available at: *http://english.people.com.cn/200507/19/eng20050719_197028.html.*

People's Daily (2005b) 'China on-campus student population exceeds 20 million'; available at: *http://english.people.com.cn/200501/27/eng20050127_172127.html.*

People's Daily (2005c) 'Intel sues Dongjing' (in Chinese); available at: *http://it.people.com.cn/GB/8219/43795/43798/3145757.html.*

People's Daily (2005d) 'Perspectives on trial no. 1 of the Chinese IT industry' (in Chinese); available at: *www.people.com.cn/GB/paper49/14393/1280057.html.*

Perrow, C. (1970) *Organisational Analysis: A Sociological View*. London: Tavistock.

Pfeiffer, P. (2000) *Sicherung von F&E-Kompetenz in multinationalen Pharmaunternehmen: Marktdruck und Technologiesprünge als Treiber neuer Organisationsformen*. Bamberg: Difo.

Robertson, R. (1995) 'Glocalization: time-space and homogeneity-heterogeneity', in Featherstone, M., Lash, S. and Robertson, R. (eds) *Global Modernities*. London: Sage Publications.

Roessner, D., Alies, C.P., Feller, I. and Parker, L. (1998) 'How industry benefits from NSF's engineering research centers', *Research Technology Management*, 41(5): 40–4.

Ronstadt, R.C. (1977) *Research and Development Abroad by U.S. Multinationals*. New York: Praeger.

Ronstadt, R.C. (1978) 'International R&D: the establishment and evolution of research and development abroad by seven U.S. multinationals', *Journal of International Business Studies*, 9: 7–24.

Schumpeter, J.A. (1950) *Kapitalismus, Sozialismus und Demokratie*, 2nd edn. Berne: A. Francke AG Verlag.

Smilor, R.W. (1987) 'Commercializing technology through new business incubators', *Research Management*, September/October: 36–41.

Smith, P.B., Peterson, M.F. and Wang, Z.M. (1996) 'The manager as mediator of alternative meanings: a pilot study from China, the U.S. and U.K.', *Journal of International Business Studies*, 27(1): 115–37.

Spark, J.D. (1985) 'The creative connection: university-industry relations', *Research Management*, 28(6): 19–21.

State Intellectual Property Office (2005a) 'Speech of Vice Premier of the State Council Wu Yi at the China-US IPR roundtable'; available at: *http://ipr2.mofcom.gov.cn/aarticle/speechactivity/200506/20050600134741.html.*

State Intellectual Property Office (2005b) 'Main responsibilities'; available at: *http://ipr2.mofcom.gov.cn/aarticle/aboutus/200506/20050600130058.html.*

Tang, Y.C. and Tian, S.H. (2005) 'Thoughts on improving the US full text patent archival system at SIPO' (in Chinese); available at: *www.sipo.gov.cn/sipo/wxfw/wxyjyfz/zlwxyj/2005/t20050812_52332.htm.*

The Economist (1999) 'Testing GM's shock absorbers', *The Economist*, 1 May.

The Economist (2004a) 'The halo effect'; available at: *www.economist.com/displaystory.cfm?story_id=321943.*

The Economist (2004b) 'Time to hit the breaks'; available at: *www.economist.com/displaystory.cfm?story_id=2668182.*

The Economist (2005a) 'China and world economy – from T-shirts to T-bonds'; available at: *www.economist.com/displaystory.cfm?story_id=4221685.*

The Economist (2005b) 'China's internet – game on'; available at: *www.economist.com/displaystory.cfm?story_id=3726963.*

The Economist (2005c) 'China's people problem'; available at: *www.economist.com/PrinterFriendly.cfm?story_id=3868539.*

The Economist Intelligence Unit (2004) *World Investment Prospects – The Revival of Globalisation?* London, New York and Hong Kong: The Economist Intelligence Unit.

Triandis, H.C. (1995) *Individualism and Collectivism.* San Francisco: Westview Press.

Tseng, H.C. (1990) 'A study of the key factors influencing the technological cooperation between firms and universities', master's thesis, Chengchi University, Taipei, unpublished.

UNCTAD (2000) *Tax Incentives and Foreign Direct Investment – A Global Survey*. New York and Geneva: United Nations.

UNCTAD (2005) *World Investment Report 2005 – Transnational Corporations and the Internationalization of R&D*. New York and Geneva: United Nations.

US Trade Representative Office (2005) 'Special 301 report of United States Trade Representative'; available at: *www.ustr.gov/assets/Document_Library/Reports_Publications/ 2005/2005_Special_301/asset_upload_file835_7647.pdf*.

Vaario, J. (2005) 'Bridging the gap between university labs and the market', paper presented at Conference for Corporate Innovation and R&D Management, Shanghai, 7–8 June, unpublished.

von Zedtwitz, M. (1999) *Managing Interfaces in International R&D*. Bamberg: Difo.

von Zedtwitz, M. (2004) 'Managing foreign R&D laboratories in China', *R&D Management*, 34(4): 439–52.

von Zedtwitz, M. (2005) 'International R&D strategies in companies from developing countries – the case of China', invited paper presented at UNCTAD Expert Meeting on Impact of FDI on Development: Globalization of R&D by Transnational Corporations and Implications for Developing Countries, Geneva, 24–27 January, unpublished.

von Zedtwitz, M. and Gassmann, O. (2002) 'Market versus technology drive in R&D internationalization: four different patterns of managing research and development', *Research Policy*, 31: 569–88.

Walsh, K. (2003) *Foreign High-tech R&D in China – Risks, Rewards, and Implications for US-China Relations*. Washington, DC: Henry L. Stimson Center.

Walsh, K. (2004) 'Testimony before the US-China Economic & Security Review Commission Hearing on "China's high-technology development"'; available at: *www.uscc.gov/hearings/2005hearings/written_testimonies/05_04_14wrts/berman_kenneth_wrts.pdf.*

Weinstein, V. and Fernandez, D. (2004) 'Recent developments in China's intellectual property laws', *Chinese Journal of International Law*, 3(1): 227–40.

Westney, D.E. (1993) 'Cross-Pacific internationalization of R&D by U.S. and Japanese firms', *R&D Management*, 23(2): 171–81.

Wharton (2005a) 'BCG-K@W Report: Overcoming the challenges in China operations – R&D in China: the next frontier or a sure bet to squander intellectual property?'; available at: *http://knowledge.wharton.upenn.edu/index.cfm?fa=viewArticle &id=1165.*

Wharton (2005b) 'BCG-K@W Report: Overcoming the challenges in China operation – the changing face of management in China'; available at: *http://knowledge.wharton.upenn.edu/index.cfm?fa=viewArticle&id=1164.*

Wingrove, N. (1995) 'China traditions oppose war on IP piracy', *Research Technology Management*, 38(3): 6–7.

World Bank (2002) 'World development indicators database'; available at: *www.worldbank.org/data/countrydata/countrydata.html.*

Wortmann, M. (1990) 'Multinationals and the internationalization of R&D: new developments in German companies', *Research Policy*, 19(2): 175–83.

Wu, F.H. (2001) 'University-industry research cooperation – a study of the major factors influencing performance', working paper, Graduate Institute of Technology and Innovation Management, Chengchi University; available at: *www.apec-stprc.org/Reports/NIS/99P0201.pdf.*

Wu, Y. (2000) 'A survey on foreign-established R&D units in China' (in Chinese), *China's Soft Science*, 1: 64–6.

Xie, W. and White, S. (2004) 'Sequential learning in a Chinese spin-off: the case of Lenovo Group Limited', *R&D Management*, 34(4): 407–22.

Xinhua News Agency (2001) 'More multinationals locate R&D centres in Shanghai', 13 December; available at: *www.xinhuanet.com/english/*.

Xinhua News Agency (2003) 'Shanghai announces R&D centre construction plan', 10 February; available at: *www.xinhuanet.com/english/*.

Xinhua News Agency (2005) 'Economic observation: "R&D base of MNCs" – when it comes to China?' (in Chinese); available at: *www.xinhuanet.com/newscenter/gnzxbb.htm*.

Xu, H.F. (2004) *Legal Practice for Patent Protection in Enterprise*. Beijing: China Mechanical Engineering Press.

Xue, L. and Wang, S. (2001) 'Globalization of R&D by multinational corporations in China: an empirical analysis', EAP Report Memorandum No. 01–06, National Science Foundation, Tokyo.

Yang, D. (2003a) *Intellectual Property and Doing Business in China*. London: Pergamon.

Yang, D. (2003b) 'The development of intellectual property in China', *World Patent Information*, 25: 131–42.

ZDNet Asia (2003) 'Foreign investment flunks R&D in China'; available at: *www.zdnetasia.com/news/business/0,39044229,39158792,00.htm*.

Zhang, J.W. (2005) 'Globalization of R&D – a case study from Kodak China', paper presented at Conference for Corporate Innovation and R&D Management, Shanghai, 7–8 June, unpublished.

Zhou, J. (2005) 'Tough employment fight ahead for graduates'; available at: *http://esperanto.china.org.cn/english/2005/Nov/147917.htm.*

Zweig, D. (2005) *Learning to Compete: China's Strategies to Create a 'Reverse Brain Drain'*. Hong Kong: Center on China's Transnational Relations.

863 Programme (2004) 'Annual report of 863 Programme (1997, 1998, 1999, 2000, 2001, 2002)' (in Chinese); available at: *www.863.org.cn/863_105/annals/index.html.*

973 Programme (2004) 'National key basic research development programme introduction' (in Chinese); available at: *www.973.gov.cn/973.aspx.*

Further reading

Allen, T. and Katz, R. (1995) 'The project-oriented career: a dilemma for human resource management', *R&D Management*, 25(2): 129–40.

Bayona, C., Garcia Marco, T. and Huerta, E. (2003) 'Collaboration in R&D with universities and research centers: an empirical study of Spanish firms', *R&D Management*, 32(4): 321–41.

Bentz, M.V. and Shapiro, J.J. (1998) *Mindful Inquiry in Social Research*. Thousand Oaks, CA: Sage Publications.

Berson, Y. and Linton, J. (2005) 'An examination of the relationships between leadership style, quality, and employee satisfaction in R&D versus administrative environments', *R&D Management*, 35(1): 51–60.

Björkman, I. and Lu, Y. (1997) 'Human resource management practices in foreign invested enterprises in China: what has been learned?', in Stewart, S. and Carver, A. (eds) *Advances in Chinese Industrial Studies*, Vol. 5. Greenwich: JAI Press, pp. 155–72.

Bommer, M. and Jalajas, D.S. (1999) 'The threat of organizational downsizing on the innovative propensity of R&D professionals', *R&D Management*, 29(1): 27–34.

Borgonjon, J. and Vanhonacker, W.R. (1994) 'Management training and education in the People's Republic of China',

International Journal of Human Resource Management, 5(2): 327–56.

Bosworth, D. and Yang, D. (2000) 'Intellectual property law, technology flow and licensing opportunities in the People's Republic of China', *International Business Review*, 9(4): 453–78.

Bowden, V. (1995) 'Managing to make a difference: a study of career diversity among men and women research scientists', PhD thesis, University of Manchester, unpublished.

Braun, W.H. and Warner, M. (2002) 'Strategic human resource management in Western multinationals in China: the differentiation of practices across different ownership forms', *Personnel Review*, 31(5): 553–79.

Brewer, J.D. (2000) *Ethnography*. Milton Keynes: Open University Press.

Browning, T. and Wang, C. (2004) 'Ten years of enforcement in China', *Managing Intellectual Property*, China IP Focus: 37–42.

Caloghirou, Y., Ioannides, S. and Vonortas, N. (2000) 'Research joint ventures', *Journal of Economic Surveys*, 17(4): 541–70.

Chen, J.W. (2001) 'China protects IC design', *China Business Review*, 28(6): 22–7.

Child, J. (1994) *Management in China During the Age of Reform*. Cambridge: Cambridge University Press.

Clark, D. (2002) 'The WTO's limited role in enforcement', *Managing Intellectual Property*, 115: 27–31.

Clark, D. (2004) 'Intellectual property litigation in China', *China Business Review*, 31(6): 25–9.

Cockburn, I.M., Henderson, R.M. and Stern, S. (2000) 'Untangling the origins of competitive advantage', *Strategic Management Journal*, 21(10/11): 1123–45.

Dane, F.C. (1990) *Research Methods*. Belmont: Brooks/Cole.

Das, T.K. and Teng, B.S. (2000) 'A resource-based theory of strategic alliances', *Journal of Management*, 26(1): 31–61.

Delios, A. and Björkman, I. (2000) 'Expatriate staffing in foreign subsidiaries of Japanese multinational corporations in the PRC and the United States', *International Journal of Human Resource Management*, 11(2): 278–93.

Dietz, M.C. and Elton, J.J. (2004) 'Getting more from intellectual property', *McKinsey Quarterly*, 4: 1–3.

Dietz, M.C., Shao, S., Lin, T. and Yang, L. (2005) 'Protecting intellectual property in China', *McKinsey Quarterly*, 3: 1–3.

Ding, D.Z., Ge, G. and Warner, M. (2004) 'Evolution of organizational governance and human resource management in China's township and village enterprises', *International Journal of Human Resource Management*, 15(4/5): 834–52.

Ding, D.Z., Goodall, K. and Warner, M. (2000) 'Beyond the iron rice bowl: whither Chinese HRM?', *International Journal of Human Resource Management*, 11(2): 217–36.

Emory, C.W. and Cooper, D.R. (1991) *Business Research Methods*, 4th edn. Boston, MA: Homewood, Irwin.

Gamble, J. (2000) 'Localizing management in foreign invested enterprises in China: practical, cultural, and strategic perspectives', *International Journal of Human Resource Management*, 11(5): 883–903.

Gassmann, O. and von Zedtwitz, M. (2003) 'Organizing virtual R&D teams: towards a contingency approach', *R&D Management*, 33(3): 243–62.

Goldberg, M.D. and Feder, J.M. (1991) 'China's intellectual property legislation', *China Business Review*, 18(5): 8–12.

Gong, T.D. (2005) 'Pharma moves ahead cautiously in China', *Science*, 29 July: 735–6.

Gupta, A. and Singhal, A. (1993) 'Managing human resources for innovation and creativity', *Research Technology Management*, 36(3): 42–8.

Hagedoorn, J., Link, A. and Vonortas, N. (2000) 'Research partnerships', *Research Policy*, 29: 567–86.

Han, Z. and Chong, L.C. (2005) 'Dynamic capabilities in R&D management – how to identify?', paper presented at 27th R&D Management Association Conference, Pisa, 6–8 July, unpublished.

Hausschildt, J. and Kirchmann, E. (2001) 'Teamwork for innovation – the "troika" of promoters', *R&D Management*, 31(1): 41–9.

Healey, M. and Rawlinson, M.B. (1993) 'Interviewing business owners and managers: a review of methods and techniques', *Geoforum*, 24(3): 339–55.

Holti, R., Neumann, J. and Standing, H. (1995) *Change Everything at Once: The Tavistock Institute's Guide to Developing Teamwork in Manufacturing*. London: Management Books 2000.

Jones, S. (1997) 'Localization threat forces expats into rearguard action', *China Staff*, October: 6–9.

Kim, Y.B. and Cha, J.S. (2000) 'Career orientations of R&D professionals in Korea', *R&D Management*, 30(2): 121–37.

Kim, Y.B., Min, B.W. and Cha, J.S. (1999) 'The roles of R&D team leaders in Korea: a contingent approach', *R&D Management*, 29(2): 153–65.

Law, K.S., Tse, D.K. and Zhou, N. (2003) 'Does human resource management matter in a transitional economy?

China as an example', *Journal of International Business Studies*, 34(3): 255–66.

Law, K.S., Wong, C.S. and Wang, K.D. (2004) 'An empirical test of the model on managing the localization of human resources in the People's Republic of China', *International Journal of Human Resource Management*, 15(4/5): 635–48.

Leung, K. and Kwong, J.Y.Y. (2003) 'Human resource management practices in international joint ventures in mainland China: a justice analysis', *Human Resource Management Review*, 13(1): 85–106.

Leung, K., Wang, Z. and Smith, P.B. (2001) 'Job attitudes and organizational justice in joint venture hotels in China: the role of expatriate managers', *International Journal of Human Resource Management*, 12(6): 926–45.

Li, Z.Y. (2002) *The Protection and Industrialization of Hi-tech Technology*. Beijing: Huaxia Press.

Liu, J.W. (2003) *Comparing China's IP System with the Standard of TRIPS*. Beijing: China People's Press.

Lu, X. and Perry, E. (1997) *Danwei: The Changing Chinese Workplace in Historical and Comparative Perspective*. London: M.E. Sharpe.

Lu, Y. and Björkman, I. (1997) 'HRM practices in China-Western joint ventures: MNC standardization versus localization', *International Journal of Human Resource Management*, 8(5): 614–28.

Lu, Y. and Björkman, I. (1998) 'Human resource management in international joint ventures in China', *Journal of General Management*, 23(4): 63–79.

May, T. (1997) *Social Research: Issues, Methods and Processes*, 2nd edn. Milton Keynes: Open University Press.

McMeekin, A. and Coombs, R. (1999) 'Human resource management and the motivation of technical professionals', *International Journal of Innovation Management*, 3(1): 1–26.

Micronas (2005) 'Company profile'; available at: *www.micronas.com/company/profile/index.php*.

Ng, Y.C. and Siu, N.Y.M. (2004) 'Training and enterprise performance in transition: evidence from China', *International Journal of Human Resource Management*, 15(4/5): 878–94.

Nolan, P. (1995) *China's Rise, Russia's Fall*. New York: St Martin's Press.

Nolan, P. (2001) *China and the Global Business Revolution*. New York: St Martin's Press.

Randle, K. and Rainnie, A. (1995) 'Managing creativity, maintaining control: a study in pharmaceutical research', *Human Resource Management Journal*, 7(2): 32–46.

Richards, T. (1988) *Creativity at Work*. Aldershot: Gower.

Robson, C. (1993) *Real World Research*. Oxford: Blackwell.

Rovai, S. (2005) 'HRM practices in foreign MNCs operating in the PRC: a new hybrid form?', *International Journal of Human Resource Development & Management*, 5(3): 284–304.

Sauders, M., Lewis, P. and Thornhill, A. (1997) *Research Methods for Business Students*. London: Pitman Publishing.

Schlesinger, M.N. (1997) 'Intellectual property law in China: Part I – complying with TRIPs requirements', *East Asian Executive Reports*, 19(1): 9–22.

Scott, W.R. (2002) 'The changing world of Chinese enterprise: an institutional perspective', in Tsui, A.S. and Lau, C.-M. (eds) *The Management of Enterprises in the*

People's Republic of China. London: Kluwer Academic, pp. 59–78.

Selmer, J. (2001) 'Psychological barriers to adjustment and how they affect coping strategies: Western business expatriates in China', *International Journal of Human Resource Management*, 12(2): 151–65.

Selmer, J. (2004) 'Expatriates' hesitation and the localization of Western business operations in China', *International Journal of Human Resource Management*, 15(6): 1094–1107.

Selmer, J. (2004) 'Psychological barriers to adjustment of Western business expatriates in China: newcomers vs long stayers', *International Journal of Human Resource Management*, 15(4/5): 794–813.

Selmer, J., Ebrahimi, B.P. and Li, M. (2000) 'Personal characteristics and adjustment of Chinese mainland business expatriates in Hong Kong', *International Journal of Human Resource Management*, 11(2): 237–50.

Shen, J. (2004) 'International performance appraisals: policies, practices and determinants in the case of Chinese multinational companies', *International Journal of Manpower*, 25(6): 547–63.

Shen, J. and Edwards, V. (2004) 'Recruitment and selection in Chinese MNEs', *International Journal of Human Resource Management*, 15(4/5): 814–35.

Snow, C.C. and Thomas, J.B. (1994) 'Field research methods in strategic management: contributions to theory-building and testing', *Journal of Management Studies*, 31(4): 457–80.

Tanur, J.M. (1982) 'Advances in methods for large-scale surveys and experiments', in McAdams, R., Smelser, N.J. and Treiman, D.J. (eds) *Behavioural and Social Science*

Research: A National Resource, Part II. Washington, DC: National Academy Press.

Thamhain, H.J. (2003) 'Managing innovative R&D teams', *R&D Management*, 33(3): 297–311.

The Economist Intelligence Unit (1997) 'Local heroes', *Business China*, 9 June: 1–3.

Tidd, J., Bessant, J. and Pavitt, K. (1997) *Managing Innovation*. Chichester: John Wiley & Sons.

Tsang, E.W.K. (1994) 'Human resource management problems in Sino-foreign joint ventures', *International Journal of Manpower*, 15(9): 4–21.

Twiss, B. (1992) *Managing Technological Innovation*. London: Longman.

US Department of Commerce (2000) 'International science and technology – policies, programs and investments'; available at: *www.technology.gov/Reports/itsw/SciTech2000.pdf*.

Venter, K. (2003) 'Building on formal education: employers' approaches to the training and development of new recruits in the People's Republic of China', *International Journal of Training and Development*, 7(3): 186–202.

von Zedtwitz, M. (2003) 'Initial directors of international R&D laboratories', *R&D Management*, 33(4): 377–93.

von Zedtwitz, M., Herstatt, C. and de Vries, D. (2005) 'How smaller companies internationalize R&D – results from an empirical study in China', paper presented to 27th R&D Management Association Conference, Pisa, 6–8 July, unpublished.

Wang, C.W., Wu, J.J. and Horng, R.Y. (1999) 'Creativity thinking ability, cognitive type and R&D performance', *R&D Management*, 29(3): 247–54.

Wang, X. and Kanungo, R.N. (2004) 'Nationality, social network and psychological well-being: expatriates in China', *International Journal of Human Resource*

Management, 15(4/5): 775–93.

Warner, M. (1993) 'Human resource management "with Chinese characteristics"', *International Journal of Human Resource Management*, 4(1): 45–65.

Warner, M. (1995) *The Management of Human Resources in Chinese Industry*. London and New York: Macmillan/St Martin's Press.

Warner, M. (1996) 'Chinese enterprise reform, human resources and the 1994 Labour Law', *International Journal of Human Resource Management*, 7(4): 779–96.

Warner, M. (2004) 'Human resource management in China revisited: introduction', *International Journal of Human Resource Management*, 15(4/5): 617–34.

Wass, V.J. and Wells, P.E. (1994) 'Research methods in action: an introduction', in Wass, V.J. and Wells, P.E. (eds) *Principles and Practice in Business and Management Research*. Aldershot: Dartmouth Publishing, pp. 1–34.

Wheare, H. and Clark, D. (2003) 'Litigation is becoming a real option in China for IP owners', *Managing Intellectual Property*, 133: 59–65.

Winter, S. (2003) 'Understanding dynamic capabilities', *Strategic Management Journal*, 24(10): 991–5.

Wong, C.S. and Law, K.S. (1999) 'Managing localization of human resources in the PRC: a practical model', *Journal of World Business*, 34(1): 26–40.

World Bank (2004) 'Total GDP 2004'; available at: *www.worldbank.org/data/databytopic/GDP.pdf*.

Wu, J. (2004) *Motive, Form and Function of International R&D Alliances in the Information Technology Industry in China*. Ottawa: Department of Systems and Computer Engineering, Carleton University.

Xiao, J. and Tsang, M.C. (1999) 'Human capital development in an emerging economy: the experience of Shenzhen, China', *China Quarterly*, 57: 72–114.

Yang, D. (2004) 'Intellectual property abuses: how should multinationals respond', *Long Range Planning*, 37: 459–75.

Yin, R. (1981) 'The case study crisis: some answers', *Administrative Science Quarterly*, 26(1): 58–65.

Yin, R. (1998) 'The abridged version of case study research', in Blickman, L. and Rog, D.J. (eds) *Handbook of Applied Social Research*. Thousand Oaks, CA: Sage Publications, pp. 229–59.

Young, L. (2005) 'China grants intellectual property protection for syngenta herbicide ingredient', *Chemical Week*, 167(9): 15–30.

Zhou, Z. and Huang, L.H. (2004) 'The conflict and coordination in university-corporation cooperation', *R&D Management*, 16(1): 1–4.

Zhu, C.J. (1997) 'Human resource development in China during the transition to a new economic system', *Asia Pacific Journal of Human Resources*, 35(3): 19–44.

Zollo, M. and Winter, S.G. (2002) 'Deliberate learning and the evolution of dynamic capabilities', *Organisation Science*, 13(3): 339–51.

Index

libary Sing